American Institute of Chemical Engineers

Fuels and Petrochemicals Division

Presentations at the
2007 AIChE Annual Meeting

November 4-9, 2007
Salt Lake City, Utah, USA

Printed from e-media with permission by:

Curran Associates, Inc.
57 Morehouse Lane
Red Hook, NY 12571
www.proceedings.com

ISBN: 978-1-60560-012-3

Some format issues inherent in the e-media version may also appear in this print version.

ISBN: 978-1-60560-012-3

American Institute of Chemical Engineers
Proceedings
Three Park Avenue
New York, NY 10016-5991
Phone: 212-591-8100

www.aiche.org

American Institute of Chemical Engineers

Fuels and Petrochemicals Division

TABLE OF CONTENTS

7a Fuel Cell Systems: An Introduction For The Chemical Engineer

Donald J. Chmielewski, Chemical and Biological Engineering, Illinois Institute of Technology, Suite 127, Perlstein Hall, 10 West 33rd Street, Chicago, IL 60616

This introduction to fuel cell systems is intended for anyone with a background in Chemical Engineering and aims to provide a foundation for those interested in fuel cell research. Undergraduate students are especially encouraged to attend.

17a Reactive Flash Volatilization Of Nonvolatile Carbohydrates For Synthesis Gas

Paul J. Dauenhauer, Josh Colby, and Lanny D. Schmidt. Chemical Engineering and Materials Science, University of Minnesota, 151 Amundson Hall, 421 Washington Avenue SE, Minneapolis, MN 55455

Carbohydrates such as cellulose and starch constitute a major fraction of biomass available as agricultural residue, grasses, and trees. Using $RhCe/Al_2O_3$ catalysts at residence times less than 70 milliseconds, these biopolymers can be directly converted to synthesis gas products near equilibrium. The effect of steam addition, N_2/O_2 feed ratio, and air preheat conditions were examined to maximize selectivity to synthesis gas at all considered fuel/air ratios. With the addition of the steam, ~80% of the hydrogen in glycerol can be converted directly to H_2. Selectivity to H_2 and CO from cellulose was observed ~50% near equilibrium maintaining 2/3 of the carbohydrate fuel value. Autothermal conversion is likely to occur by coupling the endothermic biomass decomposition process with highly exothermic catalytic partial oxidation. By directly contacting carbohydrate particles with a hot catalytic surface capable of providing high heating rates, the solid particle forms volatile organics which flow into a porous catalyst and reform with oxygen to form synthesis gas. Deactivation over 20 hrs of operation was not detected, possibly because the formation of char is limited by high particle heating rates.

Web Page: www.cems.umn.edu/research/schmidt/

The Suppression of Tar Evolution during Biomass Pyrolysis by Using Simple Air Treatment Method

*Weerawut Chaiwat, Isao Hasegawa, Kazuhiro Mae**

Department of Chemical Engineering, Kyoto University, Nishikyo-ku, Kyoto, 615-8510, Japan

Introduction

Biomass is recently considered as a renewable resource due to its environmental friendly characteristic. Pyrolysis is one of prominent methods to convert raw biomass to fuel gas or valuable chemicals. Most investigators have studied pyrolysis mechanism and generally recognized at least two competitive pathways for biomass pyrolysis. One is dehydration, i.e. cross-linking reaction, to form chars, tars, carbon oxides, and water. The other one involves depolymerization and volatilization to form chars and combustible volatiles[1]. The effect of pyrolysis conditions on product yields and distribution can be explained with these two competitive pathways. Primary reaction of biomass is endothermic pyrolysis which produces a large amount of tar products. These tar products have low reactivity and may cause damage in gas turbine/gas engine of biomass gasification system. In order to control the product yields and distribution of biomass pyrolysis, several thermal gas treatments at the temperature less than 400°C have been investigated to change the structure of coal and biomass and products distribution[2-8]. However, few literatures have described the effect of air on biomass treatment at low temperature.

In this paper, we presented the products distribution and analysis of precursors structure of biomass pre-treated with nitrogen or air at low temperature under various heating patterns. Then, the gas treated precursors were pyrolyzed in a flash mode to consider their different distribution of pyrolysis products. Finally, we preliminarily analyzed tar products obtained during gas treatment by both quantitative and qualitative methods.

Experimental

Samples

Japanese cedar was mainly used as a biomass sample to perform in gas treatment process. Raw biomass was crushed and sieved into 210-500 μm, then was dried in vacuum dryer prior use. Elemental composition of raw Japanese cedar consists of 50.6 wt% carbon, 5.9 wt% hydrogen, and 43.5 wt% oxygen on the dry and ash free basis. The ash content was 0.07 wt% and lignin content is approximately 38.5 wt%.

Thermo Gravimetric Analysis

In order to preliminarily investigate the influence of air treatment at low temperature, air treatment and subsequent pyrolysis processes by using thermo gravimetric analyzer (Shimadzu, TGA-50) were conducted. Biomass samples weighed about 1.5-3.0 mg were heated in an inert (He) atmosphere from room temperature to 110°C, then were kept at this temperature for 20 min to remove its moisture. The samples were continuously treated in air (He + O_2) atmosphere at a heating rate of 10 $K \cdot min^{-1}$ till 240, 280, 300, 320, 340, 380, 400°C, respectively, which are the temperatures before air gasification occurs. Then, it was switched into inert atmosphere and pyrolysed with the same heating rate until 800°C. The samples were completely combusted with air at 800°C to determine their carbon conversion. Air gasification and pyrolysis in inert gas were also performed for comparison.

Gas Pretreatment

Quartz tube reactor (inner diameter = 8.5 mm, length = 245 mm) with a temperature controller was used to conduct gas pretreatment experiment. The sample of 50-100 mg was put in a ceramic plate at the middle of the tube next to a temperature probe to avoid a temperature distribution. The sample was treated at low temperature in air and nitrogen with the flow rate of 50 $cc \cdot min^{-1}$ by heating up to 240, 280, 300, 320, 340°C, respectively, under a heating rate of 10 $K \cdot min^{-1}$ after keeping at 110°C for 20 min to remove the moisture, then it was cooled down immediately. Treated solid precursor was characterized by using Elemental Combustion System (Costech Instruments Co., ECS-4010) for ultimate analysis after dried in vacuo at 70°C for 24 h. Gas products during treatment were collected in a gas bag, and then were analyzed by a gas chromatography (Shimadzu Co., GC-14A). For the analysis of tar, acetone was used as the mobile phase, and it was fed at 0.5 mL/min to a gel permeation chromatograph (GPC) equipped with a column (Showa Denko K.K., Asahipak GF-310HQ) and a RI detector to estimate the molecular weight distribution of tar.

In order to suppress tar evolution, the condition of air treatment was changed by adjusting air concentration, air flow rate, temperature and heating rate pattern. Air treatment was conducted by heating to 300°C with different air concentration and flow rate as described in Table 1. The heating pattern was changed both by heating rate and holding time as shown in Table 2 to examine an optimum condition for tar suppression.

Table 1. Conditions of air pretreatment at 300°C with different air flow rate and concentration

Experiment	Total flow rate [cc min^{-1}]	Air flow rate [cc min^{-1}]	N_2 flow rate [cc min^{-1}]	% O_2 [% by volume]
1	50	5	45	2.1
2	50	25	25	10.5
3	50	50	0	21.0
4	40	40	0	21.0
5	10	10	0	21.0

Table 2. Conditions of air pretreatment with various heating patterns

Experiment	Treatment Tempertaure [°C]	Air flow rate [cc min^{-1}]	Heating rate [K·min^{-1}]	Treatment time
1	300	40	10	19 min
2	300	40	1	3 h 10 min
3	300	40	0.5	6 h 20 min
4	260	20	10	15 min + 1 h
5	260	20	10	15 min + 30 min
6	260	20	0.2	12 h 30 min
7	260	10	0.1	25 h

Flash Pyrolysis

Flash pyrolysis with 3,000 K·s^{-1} of heating rate was performed at 764°C by using Curie-Point Pyrolyzer or CPP (JHP-2S) to analyze pyrolysis products of gas treated and raw biomass samples. Samples weighed 0.5-1.0 mg was wrapped in different metal foils dependent on their curie points which indicate pyrolysis temperature. Gaseous products were continuously analyzed by a gas chromatography (Shimadzu Co., GC-14A) which is connected with CPP.

Results and Discussion

Effect of Air Treatment on Pyrolysis Rate

Thermo gravimetric analysis method was used to preliminarily investigate the effect of air treatment at low temperature on the pyrolysis rate. Figure 1 shows the pyrolysis profiles of air treatment at different temperatures with subsequent pyrolysis to 800°C. The gasification and pyrolysis profiles of untreated biomass are also shown for comparison. Air treatment to 240°C has almost the same pyrolysis profile as that of untreated biomass because the temperature less than 240°C is too low to occur biomass decomposition under a heating rate of 10 K·min^{-1}. By treating biomass with air over 240°C, the decomposition curve was shifted toward lower temperature with the increase in the air treatment temperature. This indicates that dehydration and/or the oxidation to CO and CO_2 may occur by the reaction with active oxygen in air. Moreover, the pyrolysis behaviors between their treatment temperatures to 500°C were obviously different. In particular, the char yields of air treatment at 300-340°C were slightly higher than that of untreated sample due to cross-linking reaction during air treatment. For the air treatment to the temperature above 350°C, the char yield after pyrolysis at 800°C was lower than that of untreated sample. It, therefore, shows that the gasification step seems to start at 350°C. From the result of TGA experiment, we can introductorily conclude that the structure of biomass was changed by cross-linking reaction with the simple method of air treatment at low temperature to obtain more char yields and gas products.

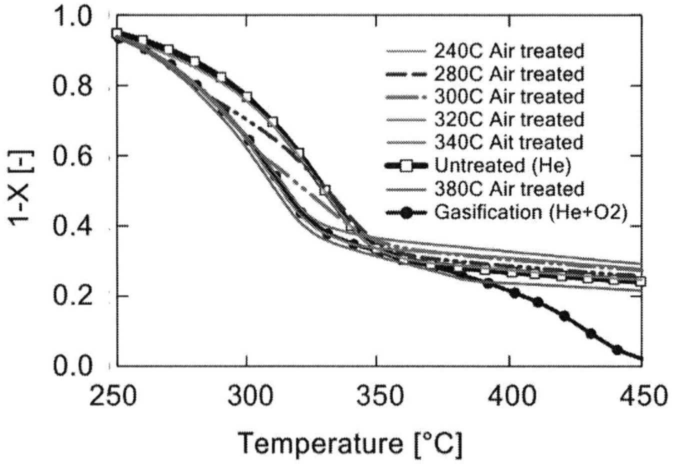

Figure 1. Weight change curves of Japanese cedar treated with air up to 240-380°C and subsequently pyrolyzed to 800°C with a heating rate of 10 K·min^{-1}.

Products Yield and Distribution during Gas Treatment and Subsequent Flash Pyrolysis

Since air treatment at low temperature has an influence on their pyrolysis rate as described in previous section, the product yield and their distribution during treatment process at various temperatures were determined. To analyze their treated solid precursors, flash pyrolysis of treated samples at 764°C was subsequently performed by using a Curie-point pyrolyzer, and the yields and distribution of pyrolysis products were compared..

(1) Influence of Air Treatment Varying with Temperature

Figures 2 and 3 show the sum of the product yields during the pretreatment and the subsequent flash pyrolysis in the same basis. The treatments were conducted in nitrogen gas and air, respectively, up to 240, 280, 300, 320, 340°C with slow heating rate (10 K·min^{-1}). The pyrolysis of each treated sample was performed in a flash mode at 764°C. The bar graphs of product yields in the figure consist of two parts: Upper part is the product yields during the pretreatment, and lower part is that during the flash pyrolysis. The yield represents based on untreated samples, so the total yield reaches more than 1 if the oxidation proceeds in the pretreatment process. During the gas treatment process, gas products such as CO_2, CO, and hydrocarbon are determined by using gas chromatography measurement, while tar and water components are calculated from the elemental mass balance of carbon and hydrogen content.

For nitrogen treatment (Figure 2), tar and gas yields increased with treatment temperature, while solid yield decreased due to the decomposition of biomass. Through different treatment temperatures, the overall yields of tar during both pretreatment and flash pyrolysis processes were almost the same as that by the untreated sample. However, char yield after flash pyrolysis slightly increased as pretreatment temperature increased because slow heating at higher temperature during pretreatment process may cause more cross-linked structure in biomass.

For air treatment (Figure 3), gas products such as CO and CO_2 clearly increased with increasing treatment temperature. Moreover, overall tar yields obtained from the air pretreatment and subsequent flash pyrolysis were reduced by ca. 10-15 wt%. Especially, at the pretreatment temperature of 300°C, tar released during air treatment was suppressed with temperature, and only little tar was produced during subsequent pyrolysis. This indicates that partial oxidation at low temperature occurred to produce CO and CO_2 while suppressing tar evolution. The structure of solid precursors treated with air at low temperature of pyrolysis was changed in a form which is difficult to release tar at higher temperature such as cross-linked structure. However, at the pretreatment temperature below 240°C, air had little effect on product yields.

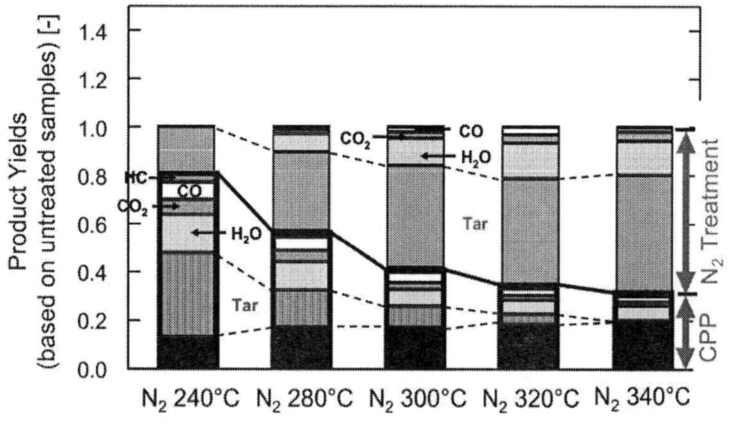

Figure 2. Product yields of Japanese cedar treated with nitrogen up to 240-340°C with a heating rate of 10 K·min^{-1} and pyrolyzed at 764°C

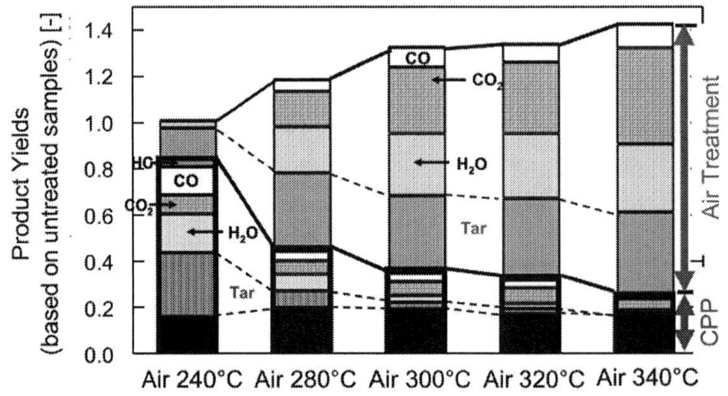

Figure 3. Product yields of Japanese cedar treated with air up to 240-340°C with a heating rate of 10 K·min^{-1} and pyrolyzed at 764°C

(2) Influence of heating pattern and holding time

In previous sections, we can decrease tar production during air treatment at low temperature, especially at 300-340°C, tar yields were reduced to about 35 wt%. However, this amount of tar can still damage gas turbine or gas engine system. Next, heating pattern of air treatment was adjusted to find a preferable condition for diminishing tar evolution.

First, 300°C treatment with air at the flow rate of 40 cc·min^{-1} and the heating rate of 1 and 0.5 K·min^{-1} was performed to compare the effect of heating rate The treatment time was then increased from 19 min to 3 hr 10 min and 6 hr 20 min, respectively. Figure 4 shows their product yields. As decreasing the heating rate, tar yields during air treatment significantly decreased to 20-30 wt% while CO, CO_2, and water much increased. Moreover, tar yield during flash pyrolysis reached to nearly zero at very low heating rates. This indicates that the structure of solid precursors treated with air contained more cross-linked bonds when gradually heating biomass at low temperature. It, thus, hardly produced tar during subsequent pyrolysis at high temperature, while CO, CO_2, and water were more produced due to aliphatic cracking of hydrocarbon by treating with air under low heating rate.

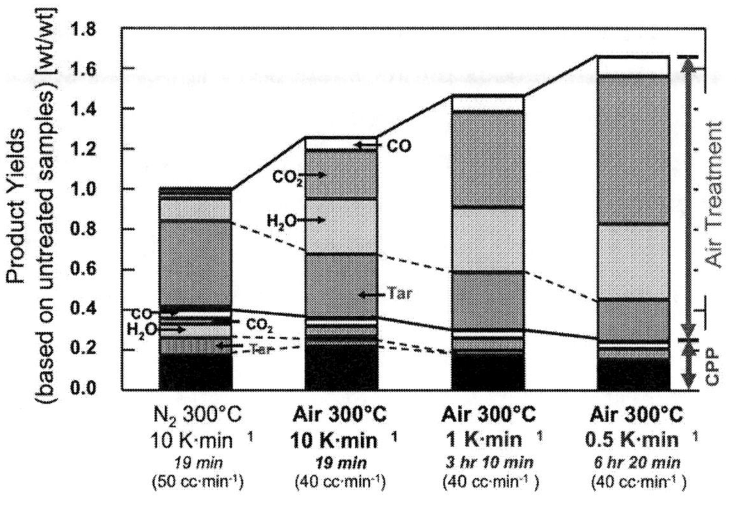

Figure 4. Product yields of Japanese cedar treated with air up to 300°C with different heating rates and pyrolyzed at 764°C

Analysis of Tar during Gas Treatment

Figure 5 summarizes the overall yield of tar produced during both gas treatment and flash pyrolysis. Overall tar yield was decreased from ca. 50 wt% to less than 20 wt% by the air pretreatment at 260°C with 0.2 K·min^{-1}. Furthermore, tar yield during flash pyrolysis at high temperature decreased through air treatment, and reached nearly zero by the air treatment at 260-300°C with heating rates of 0.2-1 K·min^{-1}.

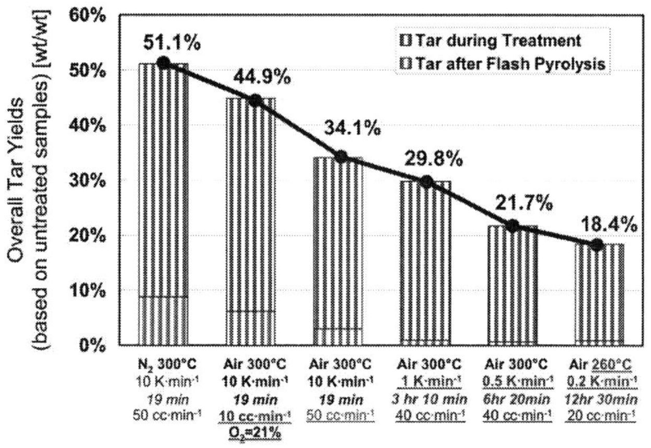

Figure 5. Overall tar yields of Japanese cedar during gas pretreatment with different heating patterns and flash pyrolysis

Tar produced during the gas pretreatment with various heating patterns was analyzed by using GPC method. Figures 6 and 7 show its molecular weight distribution. Main peaks whose retention times are in the range of 15-26 min were almost the same among the tars produced by various pretreatment conditions. As shown in Figure 7, the molecular weight distribution of tar produced by flash pyrolysis at high temperature had another different peak of high molecular weight in the range of 10-15 min, Namely, it means that tar released at low temperature about 260-300°C had relatively lower molecular weight. This result can primarily conclude that tar products obtained at low temperature during gas pretreatment may be easily decomposed by using simple reforming with steam or catalyst at low temperature.

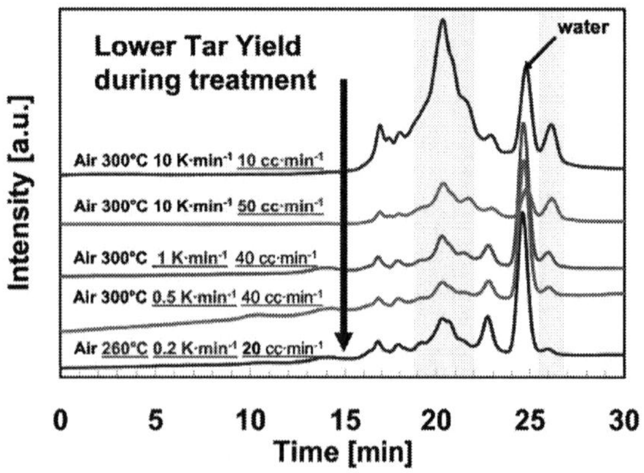

Figure 6. Molecular weight distribution of tar produced during gas pretreatment at low temperature

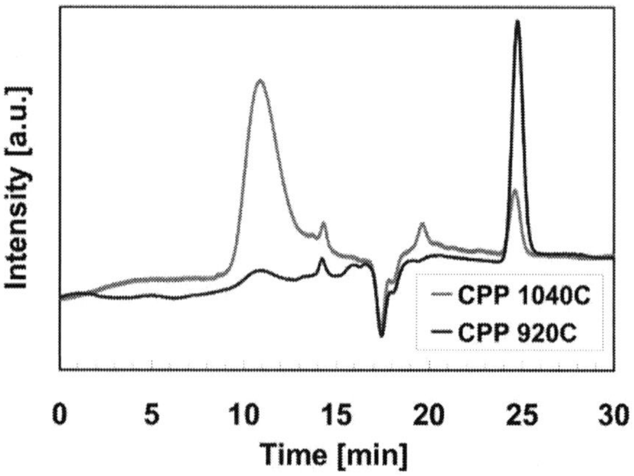

Figure 7. Molecular weight distribution of tar produced by flash pyrolysis at high temperature

Conclusions

We presented a method for reducing tar formation during pyrolysis based on the logic that the cross-linking structure brought about the decrease in tar yield. The method is a simple way that biomass is just treated by air at low temperature and then pyrolyzed at high temperature in nitrogen gas. We successfully reduced the tar yield less than several percentages by oxidizing the sample at 260-300°C with air at heating rates of 0.2-1 $K{\cdot}min^{-1}$. For tar analysis by GPC method, tar obtained during the air pretreatment at low temperature seems to be easily decomposed by steam reforming because of its lower molecular weight. This result suggests that the method can be useful as a simple method to suppress tar evolution in an on-site gasification/gas engine system for biomass utilization.

References

1. D. L. Klass (1998), "Biomass for Renewable Energy, Fuels, and Chemicals," *Academic Press*.
2. Graff, R.F., S.D. Brandes (1987), "Modification of Coal by Subcritical Steam-Pyrolysis and Extraction Yields," *Energy&Fuels*, **1**, 84-88
3. Brandes, Susan D., Robert A. Graff (1989), "Modification of Coal by Subcritical Steam: An Examination of Modified Illinois No.6 Coal," *Energy&Fuels*, **3**, 494-498
4. Miura, Kouichi, Kazuhiro Mae, Kiyoyasu Sakurada, Kanji Hashimoto (1992), "Flash Pyrolysis of Coal following Thermal Pretreatment at Low Temperature," *Energy&Fuels*, **6**, 16-21
5. Hayashi, Jun-ichiro, Yoshihiro Matsuo, Katsuki Kusakabe, Shigeharu Morooka (1995), "Effect of Light Heat Treatment on Pyrolysis Reactivity on Brown Coal," *Energy&Fuels*, **9**, 284-289
6. Zeng, Cai, Hongwei Wu, Jun-ichiro Hayashi, Chun-Zhu Li (2005), "Effects of Thermal Pretreatment in Helium on the Pyrolysis Behaviour of Loy Yang Brown Coal," *Fuel*, **84**, 1586-1592
7. Zeng, Cai, George Favas, Hongwei Wu, Alan L. Chaffee, Jun-ichiro Hayashi, Chun-Zhu Li (2006), "Effects of Pretreatment in Steam on the Pyrolysis Behaviour of Loy Yang Brown Coal," *Energy&Fuels*, **20**, 281-286
8. Worasuwannarak, Nakorn, Hiroyuki Nakagawa, Kouichi Miura (2002), "Effect of Pre-oxidation at Low Temperature on the Carbonization Behavior of Coal," *Fuel*, **81**, 1477-1484

17c Steam Reforming of Tar with Co/mgo Catalyst in Biomass Gasification Process

Osamu Hirohata, Chemical System Engineering, School of Engineering, University of Tokyo, 4-6-1 Komaba, Meguro-ku, Tokyo, Japan, Kazuhiko Tasaka, Chemical System Engineering, University of Tokyo, 7-3-1 Hongo, Bunkyo-ku, Tokyo, 113-0033, Tokyo, Japan, Chihiro Fushimi, Department of Mechamical and Biofunctional Systems, Institute of Industrial Science, The University of Tokyo, 4-6-1 Komaba, Meguro-ku, Tokyo, Japan, Takeshi Furusawa, Applied Chemystry, School of Engineering, University of Utsunomiya, 7-1-2 Yoto, Utsunomiya, Japan, Noboru Suzuki, Information and Control Systems Science, Graduate School of Engineering, University of Utsunomiya, 7-1-2 Yoto, Utsunomiya, Japan, and Atsushi Tsutsumi, Mechamical and Biofunctional Systems, Institute of Industrial Science, The University of Tokyo, 4-6-1 Komaba, Meguro-ku, Tokyo, Japan.

Biomass gasification is a promising technology for producing a fuel gas which is useful for power generation systems. During biomass gasification process, tar formation often causes some problems such as pipeline plugging. Thus, proper tar treatment is necessary. So far, Ni-based catalysts have been intensively studied for the catalytic tar removal. However, the deactivation of Ni-based catalysts takes place because of coke deposition and sintering of Ni metal particles. To overcome these problems, we have developed Co/MgO solid solution state catalyst for tar removal. The performance of Co/MgO catalysts for steam reforming of tar derived from biomass gasification in a bubbling fluidized bed gasifier of 22 mm in i.d. and 500 mm in height was investigated. These experiments were carried out by two different methods: treatment inside the gasifier with catalyst as a fluidizing medium (primary method) and hot gas cleaning in a fixed bed reactor after the fluidized bed gasifier (secondary method). Using catalysts (Co loading of 0-36 wt%) in primary method, 12 wt% Co/MgO catalyst was found to exhibit the best performance (carbon-base total tar conv. 16%) at 873 K. The results of characterizations (TPR, TPO, XRD, CO adsorption, methane pulse chemisorption, BET surface area determination, etc.) showed surface area and adsorptive capacity of tar among Co metal, Co_3O_4 and CoO-MgO caused difference of performance among catalysts with different Co loadings. The reaction mechanism of steam reforming of tar with Co/MgO catalyst was elucidated by these results. The performance using catalysts in secondary method was also measured. Comparing the results of primary method with these of secondary method, we also propose an effective process of steam reforming of tar using Co/MgO catalyst.

17d Effect of Interaction among Major Components and Alkali and Alkaline Earth Metals on Tar Evolution in Steam Gasification of Biomass

Sansan I, Chemical System Engineering, School of Engineeing,University of Tokyo, 4-6-1 Komaba Meguro-ku, Tokyo, tokyo, 153-8505, Japan, Singo Katayama, Chemical System Engineering, Institute of Industrial Science, The University of Tokyo, 4-6-1 Komaba, Meguro-ku, Tokyo, Japan, Chihiro Fushimi, Department of Mechamical and Biofunctional Systems, Institute of Industrial Science, The University of Tokyo, 4-6-1 Komaba, Meguro-ku, Tokyo, Japan, and Atsushi Tsutsumi, Chemical System Engineering, University of Tokyo, 7-3-1 Hongo, Bunkyo-ku, Tokyo, 113-0033, Tokyo, Japan.

Biomass, a carbon neutral renewable fuel, can contribute to the demand for heat, electricity and synthesis gas. Biomass gasification is a possible alternative to the direct use of fossil fuel energy. However, large amount of condensable organic compounds (tar) is evolved in biomass gasification. The presence of tar causes blockages and corrosion of pipes and also reduces overall thermal efficiency of the system. In gasification, alkali and alkaline earth metal (AAEM) species may be volatilized into the flue gas, causing corrosion of turbine blade of the gas turbine and adhesion to the wall of piping etc. as a result the heat transmission becomes worse. These inefficiencies in the technology render biomass gasification economically unviable. On the other hand, AAEM species can act as an excellent catalyst for the gasification and combustion reactions. Their retention in the char after pyrolysis would not only minimize their adverse impacts on the operation of a gasification/reforming-based power generation system, but also enhance the conversion of the solid char to gaseous products. It is, therefore, necessary to know the mechanism of tar evolution and thermal behavior of AAEM for proper design and operation of biomass gasification system. In this research, we investigated the interactions of the three main biomass components (cellulose, hemicellulose and lignin) during pyrolysis and gasification in detail by using mixtures of the three components as model biomass samples and a continuous cross-flow moving bed type differential reactor, in which tar and gases can be fractioned according to reaction time. We also investigated the effects of AAEM on the interactions among the three components by wet impregnation of AAEM into the mixtures at various temperatures. It is found that time profiles of tar evolution from the model biomass was different from those estimated from superposition of these components. This suggests the interactions between cellulose, hemicellulose and lignin in pyrolysis and gasification. It was also found that the decomposition of levoglucosan, which is the representative component of the water-soluble tar obtained from cellulose, is promoted when cellulose is mixed with lignin.

IMPROVEMENT OF BIOMASS GASIFICATION THROUGH SEVERAL PPRETREATMENTS

Isao Hasegawa, Kenichi Ikeda, Yoshikage Ohmukai, Weerawut Chaiwat, Kazuhiro Mae
Kyoto University, Kyoto-daigaku katsura, Nishikyo-ku, Kyoto, 615-8510, Japan

Introduction

As one of the thermochemical conversions of biomass, gasification has been studied extensively [1-3]. In order to achieve the efficient conversion, improvement of gasification rate and lowering of gasification temperature are required. We have developed several kinds of pretreatment methods for biomass so far. One is a method [4] for separating hemicellulose, cellulose and lignin from biomass, and for converting them into valuable chemicals. The method is basically a two-step process in which biomass treated with hot water at 180 °C was consecutively oxidized with 30%-H_2O_2 at 60 °C under ambient pressure. Through the oxidation, lignin fraction was converted to water-soluble compounds containing small molecular weight fatty acids such as HCOOH and CH_3COOH. Since there is a possibility that the aromatic structure of this water-soluble lignin became loose due to the cleavage of aromatic ring, its gasification rate is expected to be larger than that of the original lignin. From the viewpoint of the fully biomass conversion, we investigated the utilization of the water-soluble compounds as a feedstock for the gasification by comparing the gasification rates between the oxidized lignin and organosolv lignin using the so-called temperature-programmed reaction (TPR) method in a steam flow. Another pretreatment is a method of loading inorganic compounds onto biomass. A large catalytic effect in gasification is expected for some inorganic compounds. In this study, we examined the effect of these pretreatments on the gasification reactivity of biomass.

Experimental section

Samples

Paper filter cellulose (Advantec Co.) and organosolv lignin (Aldrich Co.) were used as pure biomass components. Ultimate analyses of these samples are shown in Table 1. All samples were dried in vacuo at 70 °C for 24 h prior to use.

Liquid-phase oxidation

Oxidized lignin can be separated from biomass by dissolving in aqueous solution through the liquid-phase oxidation, but the organosolv lignin was used as starting material to examine the effect of pretreatment. The liquid-phase oxidation of the organosolv lignin was performed as follows: 10 g of sample was mixed with 30% aqueous hydrogen peroxide by 1:5 of weight basis mixing ratio in a 500 mL flask with a tight plug and a gasbag. The flask with the reaction mixture was kept for 24 h in a water bath regulated at 60°C. After the elapse of the reaction time, an excess of cold water and manganese dioxide were added to the mixture to terminate the oxidation reaction, and the mixture was filtrated to be separated into the solid residue and the aqueous solution of water-soluble lignin. Dry oxidized lignin powder was obtained by evaporating the aqueous solution.

Table 1 Ultimate analysis of samples

sample	Ultimate analysys [wt%daf]		
	C	H	O(diff.)
organosolv lignin	66.4	6.0	27.6
organosolv lignin 800°Cchar	96.7	1.3	2.0
oxidized lignin	41.1	4.1	54.8
oxidized lignin 800°C char	90.7	0.0	9.3
cellulose	44.4	6.1	49.6
cellulose 700°C char	92.5	2.4	5.1
$Ca(OH)_2$-loaded cellulose 700°C char	66.7	5.3	28.0

Loading of the inorganic compound

$Ca(OH)_2$ (Wako Co.) was used as a catalyst precursor among different calcium compounds. A direct mixture of 1 g of the damped cellulose (water content: 50%) and 0.16 g of $Ca(OH)_2$ powder was carried out.

Gasification

The gasification reactivities of the sample chars were isothermally measured by use of a thermogravimetric analyzer (Shimadzu Co., TGA-50). About 2 mg of the sample mounted on a platinum cell was heated at 20 K/min up to 800 °C under a flow of pure N_2 gas, and kept for 10 minutes at 800 °C. N_2 gas was switched over to 30%-steam or 100%-CO_2 at 800 °C, and the char was gasified. In another version of the temperature-programmed-reaction (TPR) technique, sample was heated linearly in steam at ambient pressure from 150 °C, a temperature at which the reaction rate is practically zero, to 1200 °C at a rate of 10 K/min. Char conversion is expressed as weight percent on a dry ash-free basis.

In order to investigate the products during pyrolysis, flash pyrolysis was conducted using a Curie-point pyrolyzer (Japan Analytical Ind., JHP-2S). About 2 mg of sample wrapped up tightly in a ferromagnetic foil was placed in a small quartz reactor (4.0 mm i.d.) and heated to 485 °C at the rate of 3000 K/s by an induction heating coil to be pyrolyzed rapidly.

Characterization of the samples

Ultimate analyses of the samples were performed using an elemental analyzer (BEL Japan, Inc., ECS4010). Solid chemical structures such as the functional groups were analyzed by an FTIR spectrometer (JEOL. Ltd., JIR-SPX60).

Results and discussion

Effects of liquid-phase oxidation on gasification of lignin

First, we compared the gasification profiles between the oxidized lignin and the organosolv lignin in order to examine the effects of the liquid-phase oxidation on the gasification reactivity of the lignin. Figure 1 shows the TPR data of the organosolv lignin, the oxidized lignin and their chars. Two-step profiles can be seen for the organosolv and the oxidized lignin. The weight losses in the first step seem to correspond to pyrolysis and those in second step correspond to gasification of the residual chars with steam. For these lignin samples, the weight losses in first step started at around 200 °C but the weight losses in the second step started at different temperatures from each other. Gasification of the oxidized lignin proceeded at a considerably lower temperature than that of the organosolv lignin did. The oxidized lignin was so reactive that steam gasification occurred under a lower temperature region around 700 °C while the organosolv lignin was gasified above 900 °C. From this result, it was clarified that the liquid-phase oxidation was very effective for

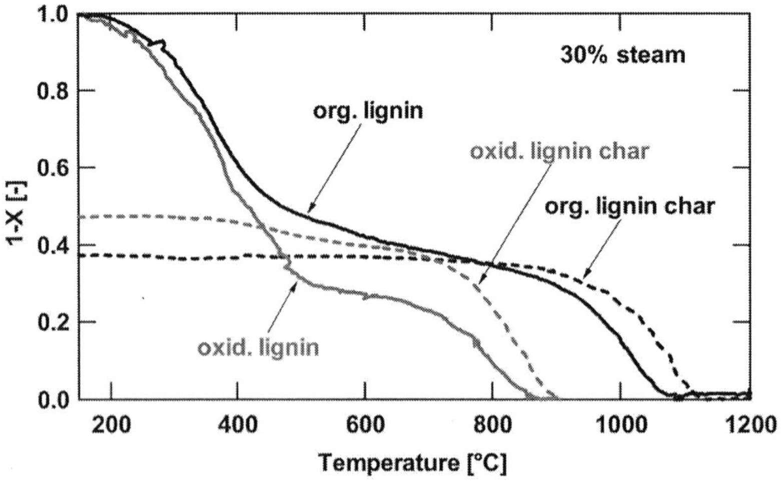

Figure 1. TPR gasification profiles of organosolv lignin and oxidized lignin and their chars

gasification of lignin as well as for separation of lignin fraction. Furthermore, even if once the lignin samples were carbonized at 800 °C, their chars also showed a similar tendency. In other words, the oxidized lignin char was gasified at a lower temperature than the organosolv lignin char was.

Thus, in order to make clear the reason of this activation effect, we investigated the differences in chemical properties among the lignin samples. Elemental compositions of the lignin samples are additionally listed in Table 1. The O/C and H/C atomic ratio values of the oxidized lignin were higher than those of the organosolv lignin. This result suggests that the chemical structure of the organosolv lignin changed into the more aliphatic one at the same time the organosolv lignin was oxidized through the liquid-phase oxidation. Next, the changes in the chemical structure of the lignin samples during pyrolysis were examined by using an FTIR spectrometer. Figure 2 shows the changes in the *in-situ* FTIR spectra for the organosolv lignin and the oxidized lignin with the increase of temperature. For the organosolv lignin, the intensity of ester groups (assigned 1800 cm^{-1}) increased during the heating. On the hand, the formations of the ester bond and the aromatic C-H bond (assigned 3050 cm^{-1}) were not observed at all for the oxidized lignin. This result indicates that most of the aromatic rings in the lignin chemical structure were cleaved through the liquid-phase oxidation. And the oxidized lignin was pyrolyzed without the formation of ester cross-linking. Judging from the above discussion, it

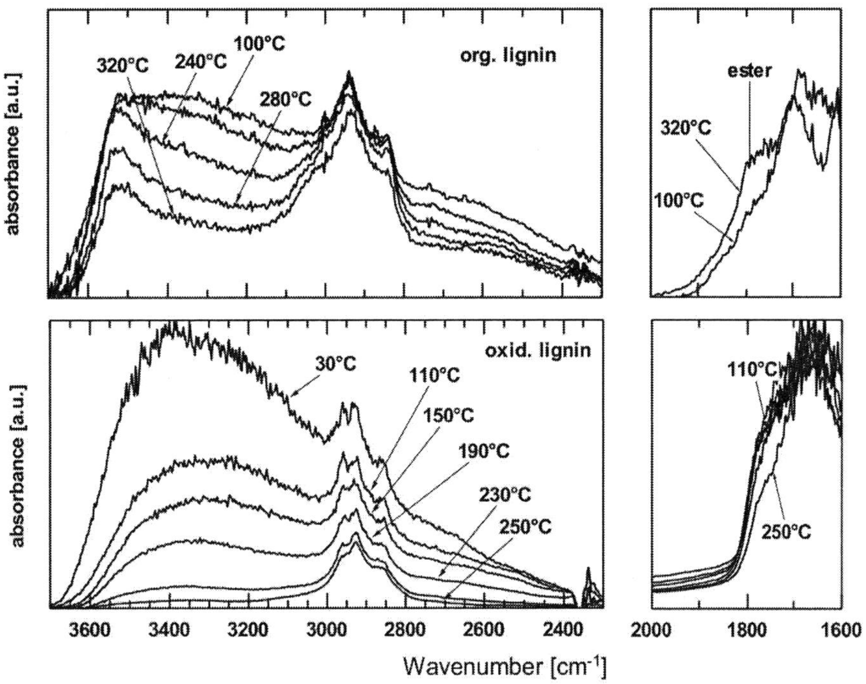

Figure 2. *in-situ* FTIR spectra of organosolv lignin and oxidized lignin during pyrolysis

was found that the activation of the chemical structure and keeping free from the inactive cross-linking during pyrolysis brought about the great improvement of gasification rate of lignin.

Effects of $Ca(OH)_2$ on pyrolysis and gasification of cellulose

Secondly, we investigated the effects of $Ca(OH)_2$ on the gasification of biomass. From the previous section, the change in chemical structure during pyrolysis at low temperature affected the gasification reactivity of the char. There is a possibility that loading of $Ca(OH)_2$ also affects the pyrolysis behavior of biomass and gasification reactivity after all. In our previous study [5], it has been clarified that the temperature range where a drastic weight loss occurred shifted to the higher temperature around 450 °C for the $Ca(OH)_2$-loaded coffee grounds. Cellulose, the main component of biomass, is decomposed into a large amount of tar during pyrolysis. It is desired that formation of tar during the heating is suppressed as much as possible. Thus, the products yields of pyrolysis were measured for pure cellulose and the $Ca(OH)_2$-loaded cellulose by using a Curie-point pyrolyzer. Figure 3 shows the product distributions of the pure cellulose and the $Ca(OH)_2$-loaded cellulose during pyrolysis at 485 °C. Tar yield for the $Ca(OH)_2$-loaded cellulose was reduced nearly to half as compared to that for the pure cellulose. On the contrary, char yield for the $Ca(OH)_2$-loaded cellulose was 4 times larger than that for the pure cellulose.

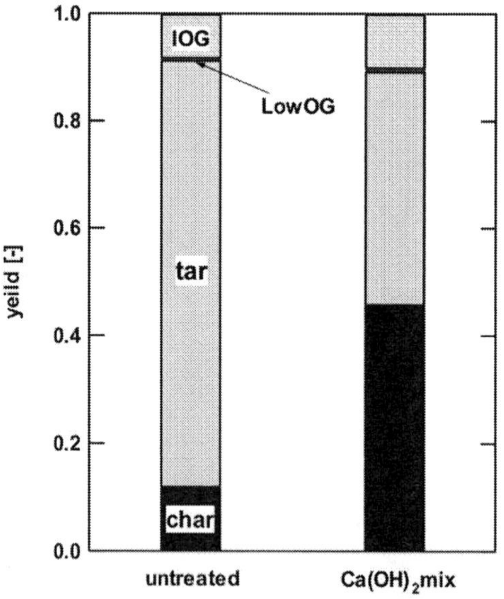

Figure 3. Comparison between products yields during 485 °C flash pyrolysis of pure cellulose and $Ca(OH)_2$-loaded cellulose

In order to examine the effect of $Ca(OH)_2$-loading on the property of the residual char, the elemental analysis was carried out. The elemental compositions of the 700 °C chars from the $Ca(OH)_2$-loaded cellulose and the pure cellulose are additionally listed in Table 1. While carbonization has progressed to a large extent (carbon content: 90%) for the pure cellulose char, hydrogen content and oxygen content still remained high for the $Ca(OH)_2$-loaded cellulose char. From the FTIR spectra of the $Ca(OH)_2$-loaded cellulose char, a large amount of remaining hydroxyl group was confirmed. These results suggest that the interaction between $Ca(OH)_2$ and the functional group induces the resistance against decomposition.

Finally, we checked whether the $Ca(OH)_2$-loaded cellulose char was more reactive in gasification than pure cellulose char or not. CO_2 and steam isothermal gasification profiles at 800 °C for the $Ca(OH)_2$-loaded cellulose char and the pure cellulose char were compared in Figure 4. For both CO_2 and steam gasification, the improvement in gasification rate were observed by loading of $Ca(OH)_2$. CO_2 gasification reaction took only 700 seconds to finish for the $Ca(OH)_2$ -loaded cellulose char, which was extremely shorter than for the pure cellulose char. $Ca(OH)_2$-loading showed the more effective catalytic effect in CO_2 gasification compared to steam gasification. This seems to be because that calcium existed as $CaCO_3$ in CO_2 gasification. And gasification rate of the $Ca(OH)_2$-loaded cellulose was still high even if calcium was removed with hydrochloric acid. This result implies that the improvement of gasification rate was due to not only the catalytic effect of calcium compounds but also the active structure of the char.

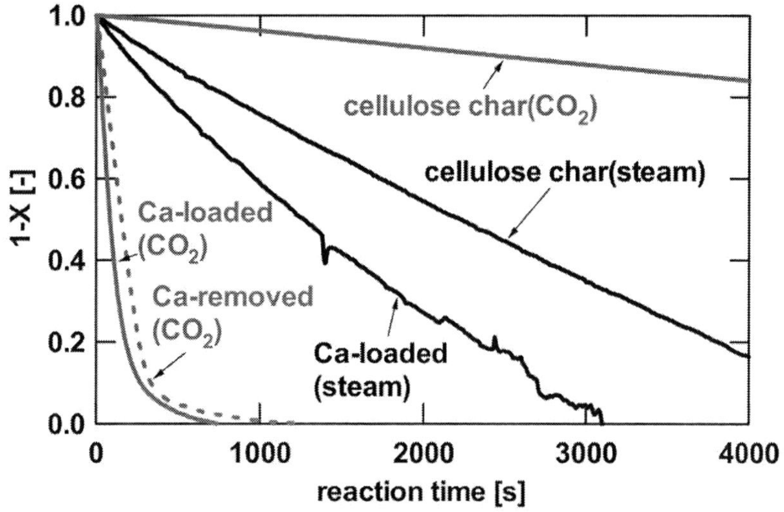

Figure 4. Isothermal gasification profiles of pure cellulose char and $Ca(OH)_2$-loaded cellulose char at 800 °C

Conclusion

Pretreatment methods for improving biomass gasification were presented. The methods are the liquid-phase oxidation and loading of $Ca(OH)_2$. The oxidized lignin was rapidly gasified at around 700 °C while the organosolv lignin was gasified above 900 °C. From the in-situ FTIR measurements, it was found that the structure of the organosolv lignin changed into cross-linked one during pyrolysis before being gasified. On the other hand, the oxidized lignin was pyrolyzed without the cross-linking reaction and developing aromatic ring structure. Loading of $Ca(OH)_2$ brought about the reduction of tar yield during the cellulose pyrolysis. A large amount of hydroxyl group remained in the residual char. By keeping reactive structure and the catalytic effect of the calcium compound, gasification rate of the $Ca(OH)_2$-loaded cellulose char increased significantly. Thus, the proposed method is very effective to improve biomass gasification.

References

1. Dalai, A. K., Sasaoka, E., Hikita, H., Ferdous, D., (2003), "Catalytic gasification of sawdust derived from various biomass", *Energy Fuels* **17**, 1456–1463
2. Elliot, D. C., Neuenschwander, G. G., Phelps, M. R., Hart, T. R., Zacher, A. H., Silva, L. J., (1999), "Chemical processing in high-pressure aqueous environments. 6. Demonstration of catalytic gasification for chemical manufacturing wastewater cleanup in industrial plants", *Ind. Eng. Chem. Res.* **38**, 879–883
3. Ohtsuka, Y., Asami, K., (1995), "Steam gasification of coals with calcium hydroxide", *Energy Fuels* **9**, 1038–1042
4. Mae, K., Hasegawa, I., Sakai, N., Miura, K., (2000), "A new conversion method for recovering valuable chemicals from oil palm shell wastes utilizing liquid-phase oxidation with H_2O_2 under mild condition", *Energy Fuels* **14**, 1212–218
5. Hasegawa I., S. Horiguchi and K. Mae (2006), "Suppression of tar formation by co-pyrolysis of calcium-loaded biomass and low temperature bio-oil," in 232nd American Chemical Society National Meeting & Exposition.

73a Production of Liquid Fuels Via Catalytic Glycerol Processing Coupled with Fischer-Tropsch Synthesis

Dante A. Simonetti[1], Edward L. Kunkes[1], Jeppe Rass-Hansen[2], Ricardo R. Soares[3], and James A. Dumesic[1]. (1) Department of Chemical and Biological Engineering, University of Wisconsin - Madison, 1415 Engineering Drive, Madison, WI 53706, (2) Department of Chemistry, Center for Sustainable and Green Chemistry--Danish Technical University, Lyngby, Denmark, (3) Faculdade de Engenharia Química, Universidade Federal de Uberlândia, Av. João Naves de Ávila 2121, Uberlandia, Brazil

We have studied the direct conversion of glycerol to liquid alkanes via an integrated process involving catalytic conversion of concentrated glycerol solutions to H_2/CO gas mixtures (synthesis gas) combined with subsequent Fischer-Tropsch synthesis in the same reactor. We show that the addition of Re to C-supported Pt-based catalysts increases the rate of synthesis gas production by an order of magnitude at conditions leading to higher CO pressures. Also, we show that this Pt-Re/C catalyst produces synthesis gas at high rates and selectivities suitable for Fischer-Tropsch synthesis (H_2/CO ratios between 1.0 and 1.6) from concentrated glycerol feed solutions (between 50 wt% and 80 wt%) at low temperatures (548 K) and high pressures (5-17 bar). Importantly, the water and oxygenated hydrocarbon byproducts from glycerol conversion either have negligible effects on the subsequent Fischer-Tropsch synthesis (*e.g.*, water, acetone, and ethanol) or have synergistic effects by participating in hydrocarbon chain growth (*e. g.*, acetol). Finally, we show that liquid hydrocarbons can be produced from glycerol in a two-bed, single reactor system at 548 K and pressures between 5 and 17 bar. The hydrocarbon products have a selectivity to C_{5+} alkanes of 75% at 17 bar, and the aqueous liquid effluent contains between 5 wt% and 15 wt% methanol, ethanol, and acetone, which can be separated from the water via distillation and used as chemicals or recycled for conversion to gaseous products. This integrated process has the potential for improving the economics of liquid fuel production from biomass via Fischer-Tropsch synthesis.

CATALYTIC GLYCEROL CONVERSION

Trung Hoang[1], Tanate Danuthai[2], Lance. L. Lobban[1], Daniel E. Resasco[1], and **Richard G. Mallinson**[1]

[1]*School of Chemical, Biological, and Materials Engineering, University of Oklahoma, 100 E. Boyd St., Norman, OK, 73019, USA*
[2]*The Petroleum and Petrochemical College, Chulalongkorn University, Soi Chula 12, Phyathai Rd., Patumwan, Bangkok, 10330, Thailand*

ABSTRACT

The catalytic conversion of glycerol by two alternative paths under various pressure and temperature conditions has been investigated. The first approach is the direct conversion of glycerol over various acidic zeolites, specifically HZSM-5, HZSM-11, HZSM22, HY, HMOR, and H-OMEGA. The second approach utilizes two consecutive catalytic beds, one with Pd/ZnO among others for deoxygenation of glycerol and one with the acidic zeolites for dehydration and oligomerization. The results show that aromatics were produced from the reaction on three dimensional pore zeolites, for instance, HZSM-5, HZSM-11, and HY zeolites. Although, the pressure was elevated to 300 psig, aromatics were still produced. However, no aromatics were produced from one dimensional pore structure zeolites (HMOR and H-OMEGA). The products observed from the reaction included propenal, acetol, acetaldehyde, propanal, propanediol, acetone, etc. Using Pd/ZnO catalysts in the first bed, the glycerol could be deoxygenated to C_1-C_4 oxygenates containing less oxygen in the molecules. Interestingly, a higher volume of a non-polar phase was obtained from the two consecutive catalytic beds (Pd/ZnO followed by HZSM-5). C_8-C_9 aromatics were yielded as the main products in that phase.

INTRODUCTION

In a period of less availability of petroleum reserves and increased awareness of environmental impact, much attention has been drawn to renewable energy sources, especially biofuels. Large amounts of glycerol are produced as a byproduct in the production of biodiesel via transesterification. Although, glycerol is currently utilized in the pharmaceutical and food industries, its volume will greatly exceed the market. In addition, glycerol from the biodiesel process is of poor quality, thus lowering its value. The objectives of the work are to utilize glycerol and to upgrade it to products that would be fungible with gasoline or for use as a chemical feedstock via catalytic conversion on zeolite catalysts.

Methanol to Gasoline (MTG) is a well known process for the conversion of methanol to hydrocarbons and water over Zeolites A, Erionite, and HZSM-5 catalysts developed by Mobil long ago. The hydrocarbons obtained in this reaction are alkanes in the gasoline boiling range (C_4-C_{10}) but with substantial amounts of aromatics [Chang et al., 1979]. Similarly, Gayubo et al. reported the successful transformation of other alcohols and oxygenates, such as, propanol, butanol, acetone, and butanone to higher paraffins and olefins over HZSM-5 catalyst [Gayubo et al., 2004]. The proposed reaction scheme involves the dehydration of oxygenates to corresponding olefins and paraffins as the first step, followed by oligomerization of these molecules. In parallel, Pestman et al. have reported the mechanism of organic acid deoxygenation on oxides such as ZnO and TiO_2 [Pestman et al., 1997]. The reaction proceeded in two steps, including a lattice oxygen atom removal by hydrogen and the deoxygenation by refilling the oxygen vacancy.

Based on these previous studies, we have investigated here the conversion of glycerol (an oxygenate containing three hydroxyl groups) by two alternative paths under various pressure and temperature conditions. The first one is the direct conversion of glycerol over various acidic zeolites, specifically HZSM-5, HZSM-11, HY, HMOR, and H-OMEGA. The second approach utilizes two consecutive catalytic beds, one with Pd/ZnO for deoxygenation of glycerol and one with the acidic zeolites for dehydration and oligomerization.

EXPERIMENTAL

HZSM-5 and H-OMEGA zeolites were supplied by SUD-CHEMIE. HY and Na-MOR zeolites were supplied by Zeolyst International. HZSM-11 was synthesized using the procedure indicated in a patent with some modification [Chu, 1973]. Na^+ ions in the Na-MOR zeolites were replaced with NH_4^+ ions by ion exchange with 20 wt% and 40% NH_4NO_3 solution at 80 °C for 10 h to obtain 20%HMOR and 40%HMOR samples, respectively. Moreover, ZnO-supported Pd catalysts (Pd/ZnO) were prepared by incipient wetness impregnation (IWI) with Palladium (II) nitrate solution (Aldrich) of the approximate concentration to obtain a metal loading of 1% Pd.

The reaction was carried out in stainless steel reactors at 300 °C and 400 °C at W/F of 0.2 and 0.5 h. The reaction took place under either atmospheric pressure or 300 psig. Prior to the reaction, the catalyst was pretreated *in situ* under flowing nitrogen for 1 h. Glycerol or ethylene glycol were fed using a syringe pump or high pressure piston pump. For the reaction with two catalytic beds, Pd/ZnO catalyst was reduced *in situ* in flowing H_2 at 400 °C for 1 h. Glycerol was fed together with H_2 at a ratio of 15:1. The liquid products were accumulated over a 3 hour interval between the samples. They were analyzed and identified using an online FID gas chromatograph equipped with a capillary INNOWAX column, and a Shimadzu GC/MS, respectively.

RESULTS AND DISCUSSION

Products obtained from the three dimensional pore zeolites.

Products from the conversion of glycerol and ethylene glycol over HZSM-5, HZSM-11, and HY catalysts are summarized in Table 1. It was found that HZSM-5 and HY produced oxygenate compounds, acetaldehyde, acrolein, and acetol from the conversion of glycerol at 300 °C. However, only aromatics were found for glycerol conversion on HZSM-5 and HY at 400 °C at atmospheric pressure. The aromatics observed from the reaction included benzene, toluene, C_8 aromatics, and C_9 aromatics. In addition, aromatic compounds were still formed when the pressure was elevated to 300 psig. This behavior was also observed on HY zeolite catalyst.

The conversion of ethylene glycol, containing two OH groups, was also investigated to understand the role of dehydration. It was found that cyclic ethers were formed from the conversion of ethylene glycol over HZSM-5 and HY at 300 °C at atmospheric pressure. Aromatics were observed only when the temperature was increased to 400 °C. Interestingly, the conversion of ethylene glycol on HZSM-11 yielded oxygenates and aromatics although the reaction temperature was at 300 °C, as shown in Table 1.

Table 1. Products Observed from HZSM-5, HZSM-11, and HY Zeolite Catalysts at Various Conditions

Catalysts	Type of feed	Pressure (psig)	Temperature (°C)	Products observed
HZSM-5	Glycerol	0	400	Oxygenates and Aromatics
		300	300	Oxygenates
		300	400	Oxygenates and Aromatics
	Ethylene glycol	0	300	Cyclic ethers
		0	400	Oxygenates and Aromatics
HZSM-11	Ethylene glycol	0	300	Oxygenates and Aromatics
HY	Glycerol	300	300	Oxygenates
		300	400	Oxygenates and Aromatics
	Ethylene glycol	0	300	Cyclic ethers
		0	400	Oxygenates and Aromatics

Products obtained from one dimensional pore zeolites.

Products produced from the conversion of glycerol and ethylene glycol over 20%HMOR, 40%HMOR, and H-OMEGA catalysts are summarized in Table 2. The results show that no aromatics were produced from these three catalysts. Only oxygenates were formed when glycerol was used as feed. However, the 20% HMOR catalyst produced cyclic ethers when ethylene glycol was the reactant. Although the Pressure was increased to 300 psig for the conversion of glycerol on H-OMEGA, the products observed were still oxygenates.

Table 2. Products Observed from 20%HMOR, 40%HMOR, and H-OMEGA Zeolite Catalysts at 300°C and 400 °C

Catalysts	Type of feed	Pressure (psig)	Temperature (°C)	Products observed
20%HMOR	Glycerol	0	300	Oxygenates
		0	400	Oxygenates
	Ethylene glycol	0	300	Cyclic ethers
		0	400	Cyclic ethers and Oxygenates
40%HMOR	Glycrol	0	400	Oxygenates
H-OMEGA	Glycerol	0	400	Oxygenates
		300	400	Oxygenates

Product Distribution of Glycerol Conversion on HZSM-5

The conversion of glycerol on HZSM-5 catalyst with various W/F at 300 psig and 400 °C was investigated. It was observed that only a polar phase was obtained at low W/F. It contained propenal (acrolein) as the main product, followed by acetol, acetaldehyde, propanal, propandiol, propenol and acetone as shown in Figure 1a. Similar product distributions are also observed at atmospheric pressure conditions.

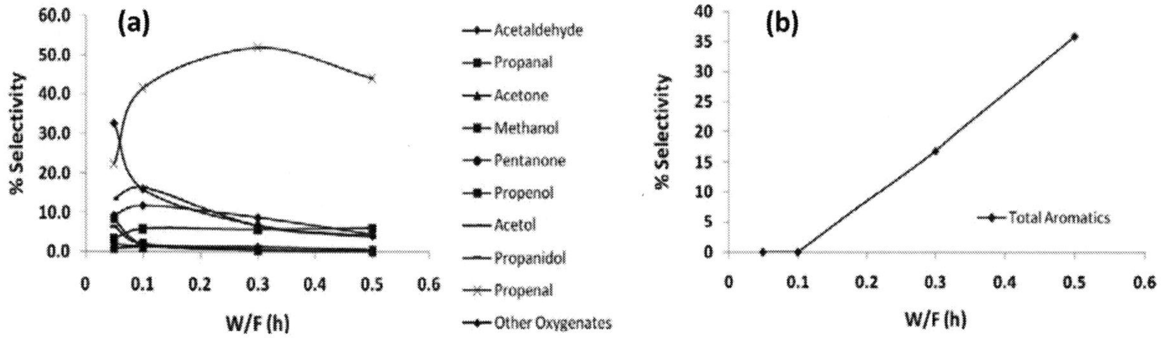

Figure 1 (a) Selectivities of oxygenates as a function of W/F over HZSM-5 (b) Selectivities of total aromatics. Conditions: 400 °C, 300 psig

When the W/F was increased to 0.3 h^{-1}, the liquid products contained two phases (polar phase and non-polar phase). The non-polar phase contains benzene, toluene, C_8 aromatics, and C_9-C_{12} aromatics. The selectivity for total aromatics increases with increasing W/F as shown in Figure 1b. Corma et al. has proposed a general reaction pathway for glycerol conversion that glycerol is dehydrated on acid catalysts to form alkanes, olefins, and oxygenates. Then, the condensation and the Diels-Alder reactions of olefins and oxygenate intermediates are involved in the production of aromatics [Corma et al., 2007].

Catalytic Activity of Glycerol Conversion on Pd/ZnO

Based on the reaction for glycerol conversion to benzene as a representative hydrocarbon, the potential (stoichiometric) volume of the hydrocarbon (non-polar phase) produced from glycerol is 45% of the total liquid products yielded from the reaction. The reaction with two catalytic beds, which were Pd/ZnO followed by HZSM-5 zeolite, were also carried out at 300 psig, 400 °C, with W/F of 0.5 h^{-1} in order to optimize hydrocarbon production. Pd/ZnO was placed in the first bed in order to deoxygenate the glycerol, followed by the HZSM-5 zeolite to oligomerize and to condense the intermediates to the final products. The products obtained from the two catalytic bed reactions are summarized in Table 3.

Table 3. Products Observed from the Conversion of Glycerol with Two Catalytic Beds at 300 psig.

Run No.	1st Bed Catalyst	2nd Bed Catalyst	Temperature (°C)	Products Observed
1	Pd/ZnO	-	400	Oxygenates
2	Pd/ZnO	HZSM-5	300	Oxygenates
3	Pd/ZnO	HZSM-5	400	Oxygenates, and Aromatics

It was found that Pd/ZnO was able to deoxygenate the glycerol to oxygenates, such as, propenal, acetol, acetaldehyde, propanal, propandiol, and acetone (Run No.1). However, it did not yield a non-polar phase. With the addition of a second bed (HZSM-5), similar products were observed from the reaction at 300 °C, compared with the reaction on Pd/ZnO alone. At 400 °C, the products from the first bed were able to react further to form aromatics (mostly C_8-C_9 aromatics). Based on these results, it can be concluded that all of the zeolites can dehydrate glycerol

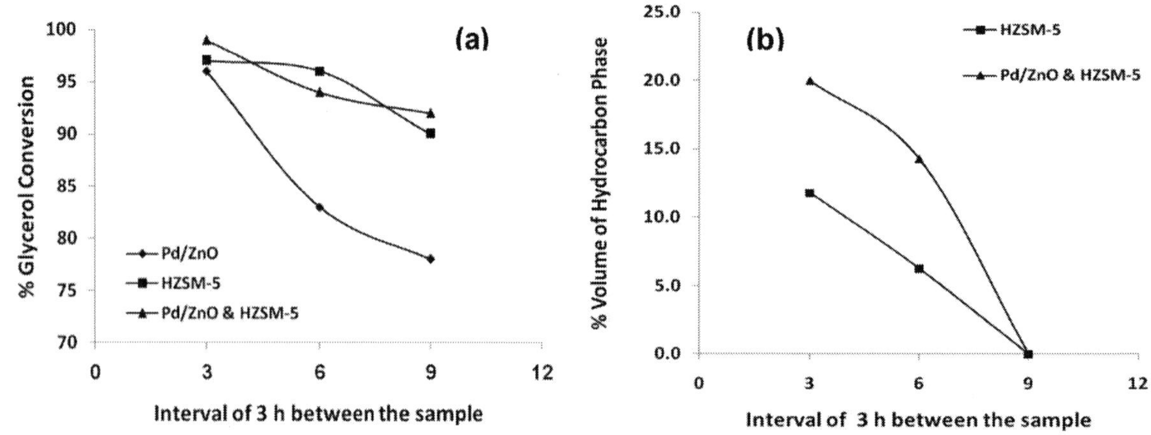

Figure 2. (a) % glycerol conversion at various intervals between the sample (b) % volume of hydrocarbon phase at various intervals between the sample.

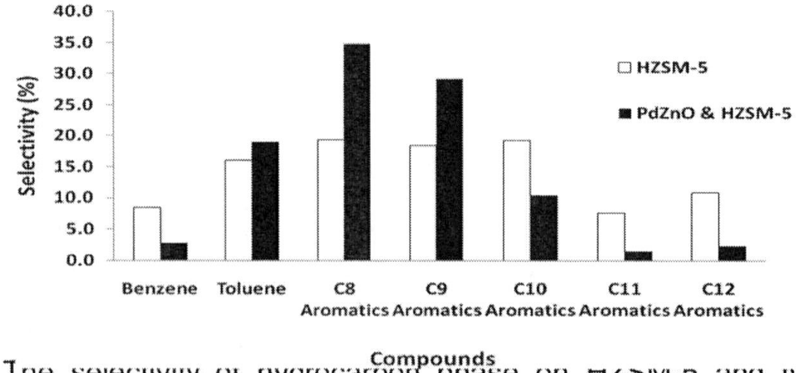

Figure 3. The selectivity of hydrocarbon phase on HZSM-5 and two consecutive catalytic beds (Pd/ZnO & HZSM-5)

The conversion of glycerol and the percentage of hydrocarbon volume obtained on Pd/ZnO, HZSM-5, and Pd/ZnO & HZSM-5 as a function of time are shown in Figure 2a and 2b. These show that glycerol conversion on HZSM-5 and the two catalytic bed reactor (Pd/ZnO & HZSM-5) did not decrease much with increasing time, compared to Pd/ZnO alone. Interestingly, a higher hydrocarbon volume was obtained from the two catalytic bed reactor. As shown in Figure 2b, the hydrocarbon phase yielded from the two catalytic bed reactor accounted for 20% by volume, while the glycerol conversion on HZSM-5 provided 12% by volume at the first 3 h. The two catalytic beds reactor still provided the higher volume of hydrocarbon up to 6 h. However, no additional aromatics were produced after 6 h. The C_8-C_9 aromatics were found as the main compounds in hydrocarbon phase, as shown in Figure 3.

CONCLUSIONS

In summary, we conclude that the structure of the zeolite affects the product distribution as follows: HZSM-5, HZSM-11 and HY produce aromatics, while no aromatics were observed from MOR and OMEGA zeolites. The acidic HZSM-5 zeolite catalyst is also able to dehydrate glycerol. Moreover, the two-catalytic-bed reaction seems to be a promising direction for the production of fungible gasoline-range molecules and chemicals. Glycerol is deoxygenated in the first Pd/ZnO bed to C_1, C_2, and C_3 oxygenates that can react further to produce in the second bed. It produced a larger volume of hydrocarbon phase compared to HZSM-5 alone.

ACKNOWLEDEMENT

The work has been financially supported by the Oklahoma Secretary of Energy. One of the authors (T.D.) would like to thank Thailand Research Fund under the Royal Golden Jubilee Ph.D. Program for scholarship as well as his mentor Professor Somchai Osuwan.

REFERENCES

Chang C.D., Lang W. H., Smith R.L., The conversion of methanol and other O-compounds to hydrocarbons over zeolite catalysts, J. Catal., 56, 169-173, 1979.

Chu P., Crystalline zeolite ZSM-11, US Pat. 3 709 979, 1973.

Corma A., Huber G. W., Sauvanaud L., O'Conner P., Processing biomass-derived oxygenates in the oil refinery Catalytic cracking (FCC) reaction pathways and role of catalyst, J. Catal., 247, 307-327, 2007.

Gayubo A.G., Aguayo A. T., Atutxa A., Aguado R., Bilbao J., Transformation of oxygenate components of biomass pyrolysis oil on a HZSM-5 zeolite. I. Alcohols and Phenols, Ind. Eng. Chem. Rev., 43, 2610-2618, 2004

Gayubo A.G., Aguayo A. T., Atutxa A., Prieto R., Bilbao J., Deactivation of a HZSM-5 zeolite catalyst in the transformation of the aqueous fraction of biomass pyrolysis oil into hydrocarbons, Energy and Fuel, 18, 1640-1647, 2004.

Pestman R., Koster R.M, Pieterse J.A.Z., Ponec V., Reaction of carboxylic acids on oxides: 1. Selective hydrogenation of acetic acid to acetaldehyde, J. Catal., 168, 255-264, 1997

Pestman R., Koster R.M, Pieterse J.A.Z., Ponec V., Reaction of carboxylic acids on oxides: 2. Bimolecular Reaction of Aliphatic Acids to Ketones, J. Catal., 168, 265-272, 1997

73c Processing Of Biomass-Derived Feedstocks In A Petroleum Refinery: Catalytic Cracking Of Biomass-Derived Feedstocks

George Huber, Chemical Engineering, University of Massachusetts-Amherst, 159 Goessmann Lab, 686 North Pleasant St., Amherst, MA 01003, Laurent Sauvanaud, Instituto de Tecnología Química, Valencia, Spain, Paul O'Connor, BIOeCON BV, Hogebrinkerweg 15e, Hoevelaken, 3871KM, Netherlands, and Avelino Corma, Instituto de Tecnología Química (UPV-CSIC), Avda. de los Naranjos s/n, Valencia, 46022, Spain.

One option for biofuel production is to use petroleum refinery technologies (i.e. catalytic cracking and hydrotreating) for biofuels production. We will discuss the feasibility of utilizing of the FCC process for biofuel and biochemical production. We have studied the reaction pathways for conversion of glycerol and sorbitol (representative biomass-derived feedstocks) over six different zeolite based catalysts. In this process, oxygen is removed from the feed as H2O, CO or CO2. Repeated dehydration and hydrogen transfer allows the production of olefins, paraffins, and coke. Aromatics (in yields up to 20 molar carbon %) are formed by Diels-alder and condensation reactions of olefins and dehydrated species. Glycerol can also be converted by FCC when co-fed with petroleum-derived products (i.e. vacuum gas-oil) without significantly altering the product selectivity. A modified FCC process can also be used to make valuable oxygenated intermediates (including acrolein, acetaldehyde) from biomass-derived feedstocks.

Conversion of Biodiesels into Normal Diesels

Phuong T. Do[1], Tanate Danuthai[2], Somchai Osuwan[2], and Daniel Resasco[1]
[1]University of Oklahoma, Oklahoma, USA
[2]The Petroleum and Petrochemical College, Chulalongkorn University, Bangkok, Thailand

Abstract

The objective of this research is to convert biodiesel into normal diesel in order to make biodiesel more fungible. The model compound representing biodiesel for this study is methyloctanoate ($C_9H_{18}O_2$, MeOct). Reactions were carried at atmospheric pressure, at 603 K, and in the flow of hydrogen. Previously, it was found that Platinum and Palladium are active in decarboxylation of MeOct to n-heptane (C_7H_{16}). It is also known that metal oxides, such as TiO_2, ZnO, ZrO_2..., are capacable of deoxygenating carboxylic acids or esters to corresponding hydrocarbons via oxygen vacancy mechanism. Therefore, three catalysts (1%Pt/Al_2O_3, 1% Pt/TiO_2 and pure TiO_2) were tested. From the activity results, several main conclusions are drawn: - Although under the reaction conditions, pure TiO_2 is partially reduced and oxygen vacancies are formed to some extent, the dominant products on TiO_2 are heavy compounds: symmetrical ketone ($C_{15}H_{30}O$) and ester ($C_{15}H_{30}O_2$), not alkanes. The catalyst deactivates fast due to strong adsorption of those heavy compounds on the catalyst surface. - When Pt is added on titania, the formation of heavy compounds is much suppressed. At the same time, both n-heptane and n-octane (C_8H_{18}) were observed. However, n-octane is the main deoxygenation product. This is due to the enhanced hydrogen spill-over effect from Pt to TiO_2 support. More oxygen vacancies are formed compared with pure TiO_2. In addition, sequential deoxygenation was seen. First, one oxygen is removed from MeOct to produce octanal (or octanol), and then these one-oxygen containing compounds deoxygenate to produce n-octane. - On alumina-supported Pt catalyst, n-heptane, not n-octane, appears as the main product. Here decarboxylation reaction on Pt is dominant compared to deoxygenation reaction via oxygen vacancy on oxide Al_2O_3. This can be explained by the fact that TiO_2 is easier to be reduced than Al_2O_3 under the operating conditions. Therefore, not many oxygen vacancies are created on Al_2O_3 support. It is also found that octanoic acid is the intermediate product in decarboxylation reaction.

I. Introduction

Biodiesels, fatty acid alkyl esters, are obtained from transferication of triglyceride molecules with short chain alcohols such as methanol or ethanol. Biodiesel has high cetane number (around 100), compared to 40 for normal diesel fuel. However, there are some problems associated with operating biodiesel in normal diesel engines. Biodiesel has low fungiblity, which is originated from oxidative and thermal instability of biodiesel, and cause operational problems at low temperatures. In attempt to make biodiesel more fungible and acceptable, many researchers have proposed deoxygenation of either triglyceride or fatty acid alkyl ester molecules to hydrocarbons, which can be used as normal diesel. Deoxygenation of triglycerides has been obtained under catalytic hydrocracking conditions by Djega-Mariadassou et al. The products included fatty acids, cyclic acids, cyclic hydrocarbons and straight paraffins and olefins. Maier et al. also reported the related work of deoxygenation of small carboxylic acids such as heptanoic and octanoic acids to produce corresponding hydrocarbons, n-hexane and n-heptane respectively. . He found that Pd/SiO_2 catalyzed decarboxylation more effectively than Ni/Al_2O_3. Recently, Murzin et al. has reported the reaction of stearic acid and ethyl stearate, which can be considered as model compounds in biodiesel, over supported noble metal catalysts. In batch condition, 5% Pd/C was the most active and selective catalyst. In the same line of deoxygenation reactions, Ponec et al. focused on reaction of acetic acid and benzoic acid with oxygen vacancy on reducible metal oxides. Deoxygenation starts with carboxylic acids reacting with oxygen vacancy to yield aldehyde, and then aldehyde further reacts with another vacancy to form hydrocarbon. The presence of Pt enhances hydrogen spill-over effect from Pt to oxides, consequently generating more oxygen vacancies. In this paper, we showed the effectiveness of various supported Pt catalysts in converting a model biodiesel compound, methyloctanoate, to corresponding hydrocarbons. The product distribution was strongly affected by using different supports.

II. Experimental

Three catalysts are used in this study. P25 titania is provided by Degussa. Alumina support is bought from Alfa Aesar and grinded into powder before use. Pt precursor is chloroplatinic acid hexahydrate from Sigma-Aldrich. Support Pt catalysts are prepared by impregnation at incipient wetness. Pt loading is 1%. Before used in reaction, catalysts are stored at 373 K for 12 hours, and then calcined at 573 K for 2 hours in air.

The catalysts are first reduced at 603 K for 2 hours before flow reactions start. The reactions are kept constantly isothermally at 603 K. Methyloctanoate and hydrogen are co-fed. Products are in vapor phase and analyzed by 6890 HP gas chromatogram. Product identification is carried by Shimazu QP5000 GC-MS and further confirmed by using standard compounds.

III. Results and Discussion

Figure I shows the reaction scheme of methyloctanoate (MeOct) on supported Pt catalysts. In hydrogen flow, two desirable products are paraffins: n-heptane (n-C7) and n-octane (n-C8). Side products include octanoic acid, octanal, octanol, di-heptyl ketone and octyloctanoate. Symmetrical di-heptyl ketone is the major product observed with TiO_2 support because TiO_2 is a well-known catalyst for ketonization reaction of carboxylic acids. Formation of undesirable ketone is strongly suppressed when Pt is added. Consequently, this gives raises for both desirable hydrocarbons (n-C7 and n-C8) on 1% Pt/TiO_2.

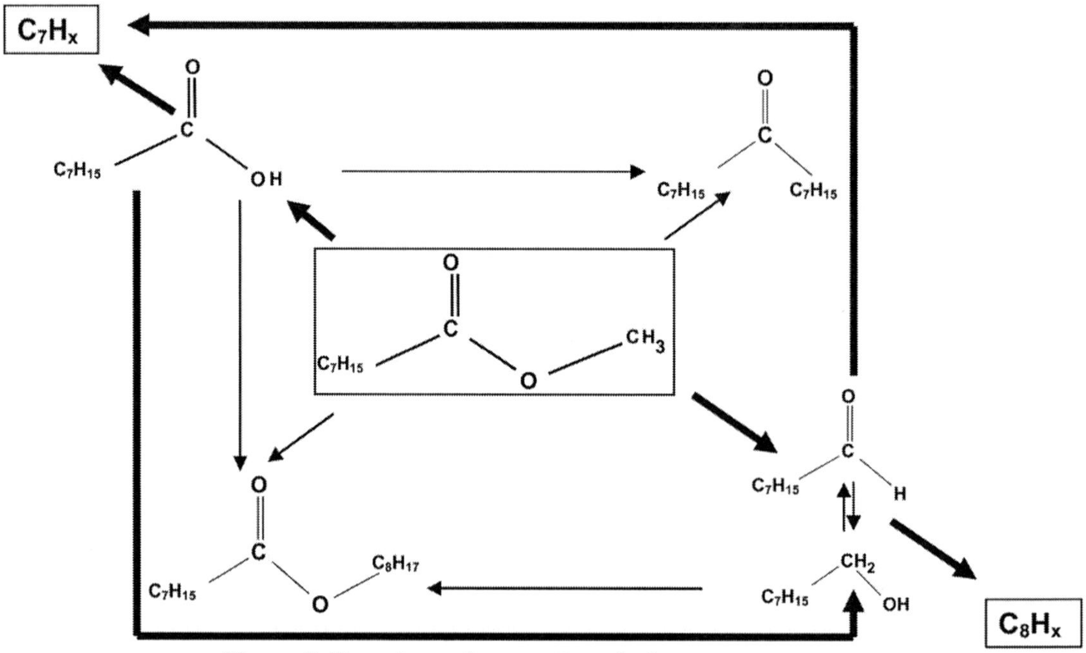

Figure I: Reaction scheme of methyloctanoate

Figure II displays the selectivity of n-C8 for three catalysts: 1%Pt/TiO_2, 1%Pt/Al_2O_3 and TiO_2. Under the same reduction condition, titania easier gets reduced than alumina. In addition, hydrogen spilled over from Pt sites to TiO_2 helps to create more oxygen vacancies on TiO_2 support, which result from loosing lattice oxygen during the reduction step of Ti cations. Similar to formation of ethane from acetic acid studied by Ponec et al., formation of n-C8 is attributed to the deoxygenation reaction of MeOct on oxygen vacancy. Therefore, Pt/TiO_2 produces more selectively n-C8 than both Pt/Al_2O_3 and TiO_2. Further analysis of selectivities of n-C8 and one-oxygen-containing compounds (octanal+octanol) indicates that octanal/octanol acts like an intermediate in deoxygenation of MeOct to n-C8. Sequential removal of oxygen from MeOct to octanal/octanol finally to n-C8 is observed.

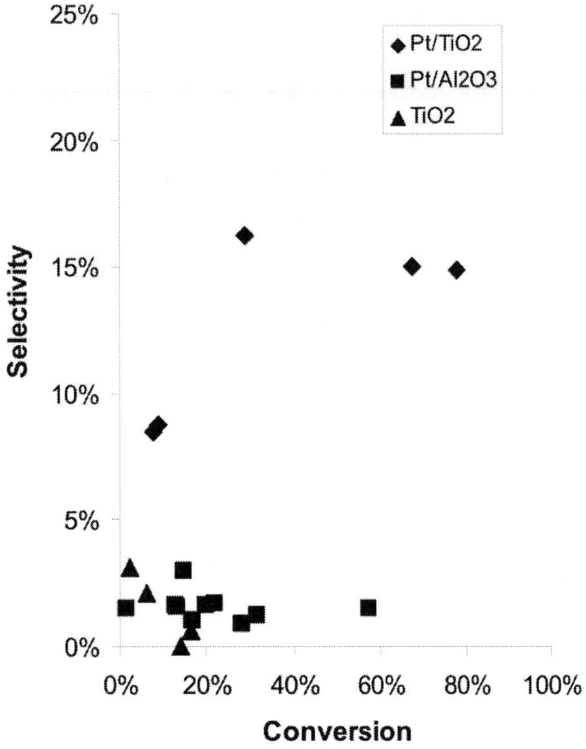

Figure II: Selectivity of n-octane over three tested catalysts

Even though Pt/TiO_2 produces more selectively n-C8 than the other catalysts, n-C7 still remains as a major product over two supported Pt catalysts. This is clearly depicted on Figure III. Selectivity of n-C7 is highest with Pt/Al_2O_3, which can be explained by the fact that only Pt, not alumina support, participates in deoxygenation reaction, which releases CO_2 from MeOct. Since alumina is considered as a non-reducible oxide, number of oxygen vacancy formed on alumina is much less than on titania. Consequently, Pt/Al_2O_3 participates in sequential deoxygenation to produce n-C8 to much less extend than Pt/TiO_2. Decarboxylation or removal of CO2 is the most selective reaction on Pt/Al_2O_3. The analysis of intermediate species on Pt/Al_2O_3 has been performed in a similar fashion as on Pt/TiO_2. n-Heptane comes from decarboxylation of octanoic acid, which results from C-O hydrogenolysis of MeOct. Both C-O hydrogenolysis and decarboxylation are postulated to occur on metal Pt.

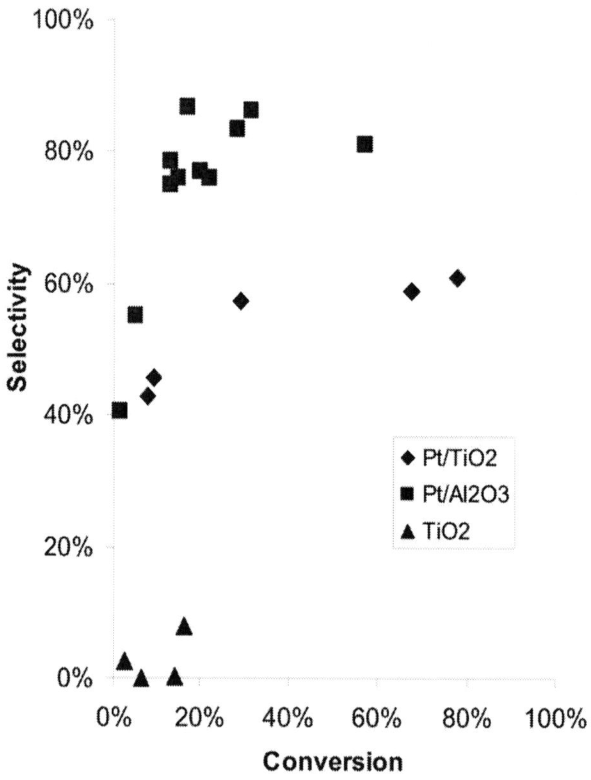

Figure III: Selectivity of n-heptane over three tested catalysts

In summary, this work has shown that support noble metal catalysts are effectively in converting a model biodiesel compound into straight-chain hydrocarbons, which can be used as normal diesel. In addition, deoxygenation reactions can occur on noble metals (decarbox/nylation) and on metal oxides (deoxygenation via oxygen vacancy). In the reaction of methyloctanoate, Pt/Al_2O_3 shows higher selectivity for n-C7 than Pt/TiO_2. In contrast, Pt/TiO_2 produces more n-C8 than Pt/Al_2O_3 due to higher ability of oxygen vacancy formation. From the understanding of different reaction mechanisms, it is possible to tailor the catalysts and operating conditions to obtain desirable product distribution (i.e. varying support, type of metals, metal loading…).

IV. References

1. "Biodiesel: The Use of Vegetable Oils and Their Derivatives as Alternative Diesel Fuels " by Gerhard Knothe, Robert O. Dunn and Marvin O. Bagby
2. Djega-Mariadassou et al. *Fuel.* 1993, 72(4), 543-549
3. Maier et al. *Chem. Ber.* , 1982, 115, 808-812
4. Murzin et al. *Cat. Tod.*, 2005, 106, 197-200
5. Murzin et al. *Ind. Eng. Chem. Res.* 2006, 45, 5708-5715
6. Ponec et al. *J of Cat.* 1997, 168, 255–264
7. Ponec et al. *J of Cat.* 1997, 168, 265–272

73e Investigation Of Catalyst Deactivation Mechanism For Hydrogen Production From Fermentation Broth

Sadashiv Swami, Chemical & Environmental Engineering, University of Toledo, 2805 Bancroft Street, 3048 Nitschke Hall, Toledo, OH 43606, Ponnaiyan Ayyappan, Department of Chemical and Environmental Engineering, University of Toledo, 2801 W. Bancroft street, Toledo, OH 43606, and Martin Abraham, University of Toledo, Dept. of Chemical Engineering, MS 310, 2801 West Bancroft, Toledo, OH 43606.

Research on hydrogen production is given much attention nowadays due to its importance as a future alternate energy source. It reduces the dependence on fossil fuel and the production of carbon-based emissions. Aqueous phase reforming of biomass and bio-derived compounds has been studied as an alternative for steam and auto-thermal reforming. However, direct conversion of biomass to hydrogen in APR is still in its infancy due to technical challenges involved in the process. Therefore, an integrated biological and thermo chemical conversion approach has been adopted. Biomass is first converted to bio-derived compounds such as ethanol which are relatively easier to reform under aqueous phase reforming conditions. APR experiments were carried out at lower temperature (225-275 °C) and moderate pressure 400-900 psi. Hydrogen yields obtained during APR of ethanol were in the range of 20-25%. Different metal and support combinations have been explored to obtain active, stable and inexpensive catalyst. While most of the catalysts were found to be stable when exposed to simulated sample (simulated sample being 5 % pure ethanol in DI water), these catalyst were rapidly deactivated when exposed to the fermentation broth. Impurities in the fermentation broth were suspected to cause catalyst deactivation. These impurities could be sulfur, phosphorus and nitrogen compounds present in cells used for the biological conversion step. An elemental analysis of the fermentation broth showed the concentration of S and P to be approximately 10 ppm. The effects of these impurities were modeled using surrogate compounds such as L-Cysteine (a sulfur-containing amino acid) and adenosine triphosphate (a phosphorus surrogate), both of which are basic ingredients of the cell structure. Catalyst deactivation in case of simulated sample doped with 10 ppm of L-cysteine was found to be of linear type. In order to eliminate the problems of catalyst deactivation, fermentation broth was treated using nano-filtration membranes. Filtered fermentation broth was subsequently used in APR process. It was observed that the catalyst performance was stable for 65 hrs of reaction time.

73f Cobalt Catalyst Development For Fisher Tropsch Synthesis

Ahmad Nafees and Saleh H. Al hashimi. Chemical Engineering Program, The Petroleum Institute, PO Box 2533, Abu Dhabi, United Arab Emirates

Commercially Fischer Tropsch (F-T) synthesis uses cobalt and iron catalyst to produce, hydrocarbons, oxygenates and chemicals including unrefined gasoline, diesel and wax ranges. Cobalt based catalyst have the advantage of higher conversion rate and a longer life. Theses catalysts are in general more reactive for hydrogenation and produce therefore less unsaturated hydrocarbons and alcohols compared to iron catalyst. The cobalt catalyst was firstly discovered in early 1920s. As soon as it was demonstrated that cobalt catalyst is the best suitable catalyst for F-T synthesis, commercial cobalt based processes were developed. From 1950s until late 1980s, iron based F-T processes dominated both industrial and R&D level area. Resulting in only marginal advancement in cobalt based process and technologies. Starting early 1990s, development of cobalt based technologies again gathered momentum.

Over the years, as a result of improvement in catalyst design technology, the performance and selectivity of the cobalt catalyst has been improved many folds. It has been demonstrated that that the activity and selectivity of cobalt based F-T synthesis catalyst can be influenced by the nature of the support (composition, pore size etc) and the type of the promoter. The idea of using composite catalyst (mixture of a classical F-T catalyst and a co-catalyst) is relatively newer. It has been shown that these catalysts have superior performance. However, the effect of physical mixing of catalyst and an adsorbent on catalytic activity and selectivity of the mixed catalysts is not well understood. One of the recent studies find that physical mixture of cobalt catalyst and an adsorbent separated by a wire mesh can significantly improve the performance. The authors are currently investigating the effect of different adsorbent parameters (catalyst/adsorbent ratio, pore size etc) on the selectivity/activity of the F-T synthesis. This manuscript reviews the historical development and present status of the cobalt catalyst for F-T synthesis.

160a Biodiesel Synthesis Using A Strong Solid Acid Catalyst

Edgar Lotero, Kaewta Suwannakarn, and James G. Goodwin Jr. Department of Chemical and Biomolecular Engineering, Clemson University, 127 Earle Hall, Clemson, SC 29634-0909

Sulfated zirconia (SZ), a solid acid with acid strength comparable to sulfuric acid, has been studied for triglyceride (TG) transesterification and free fatty acid (FFA) esterification using both model compounds and soybean oil (SBO). Even though SZ is one of the most promising solid acid catalysts for esterification of FFAs, the potential of SZ for direct transesterification of TGs remains unexplored. Under the right reaction conditions, SZ can be a suitable catalyst for the synthesis of biodiesel from low grade lipid feedstocks with high concentrations of FFAs given its ability to catalyze both esterification and transesterification. The use of a model TG compound, tricaprylin (TCP), with saturated fatty acid moieties containing eight carbons each (about half the size of a typical TG as found in SBO), allowed us to study how molecular size and chain unsaturation affected catalyst activity and catalyst deactivation. Good TCP conversion (80%) was obtained after 2 h of reaction using SZ (10 wt% based on TG content; 12:1 methanol:TG molar ratio) at 120°C. Higher reaction temperatures resulted in better reaction kinetics and conversions. However, the elevated increase in methanol vapor pressure with reaction temperature becomes a deterrent for the use of temperatures above 150°C. Under the same reaction conditions, esterification was faster than transesterification for reactions where both TGs and FFAs were present. The irreversible adsorption (and possible surface reaction) of large lipid species appears to be strongly linked to catalyst deactivation. Catalyst regeneration studies showed that calcination temperatures below 600°C are preferable to avoid the lost of surface active sulfate groups that can lead to the irreversible lost of catalyst activity.

160b Heterogeneous Catalytic Activities Of Anion Exchange Resins On The Transesterification Of Soybean Oil With Methanol

Manhoe Kim[1], John Wilson[2], Steven O. Salley[1], and K.Y. Simon Ng[1]. (1) Department of Chemical Engineering and Materials Science, Wayne State University, 5050 Anthony Wanye Drive, Detroit, MI 48202, (2) National Biofuel Energy Laboratory, NextEnergy, 461 Burroughs ST, Detroit, MI 48202

Traditional fossil energy resources are limited. One of the promising approaches is the conversion of vegetable oils. Most of the commercial biodiesel comes from the transestrification of soyoil using caustic homogeneous base catalysts. Although homogeneous processes are relatively fast, they are still not fully cost-competitive because of the purification cost and side reaction caused by caustic catalysts dissolved in the products In order to solve these problems, the development of insoluble and reusable heterogeneous catalyst is required. Anion exchange resins were tested as heterogeneous base catalysts in the transesterification of soybean oil with methanol. The catalytic activities of the resins were measured by the yield of the fatty acid methylester in the products with a GC-mass spectrometer. All of the anion-exchange resins catalyzed the transesterification and produced fatty acid methyl esters. Especially, the resins of type I (DOWEX marathon A , MSA, and SBR LCNG) having benzyl-trimethylamine functionality showed very good catalytic activities compared to the resin of type II having benzyl-dimethylethanolamine (DOWEX 22) or polyamines functionalities (Weak base anion exchanger: Marathon WBA). Transestrification reactions were completed after 13hr at 55°C when 10g of type I resin catalysts were used in the mixture of 10.8g methanol and 30.0g of soybean oil. However, 76% conversion was obtained from the weak base anion exchanger (Marathon WBA) at the same condition. A new ion exchange method using a NaOH solution (1.0M) in methanol instead of aqueous solution of NaOH was attempted in order to eliminate the negative effects of water on the deactivation of catalyst and side reactions. Catalytic activities of anion exchange resins that are different in base strength, degree of cross-linking, matrix structure, and particle sizes will be discussed.

160c Production Of Reusable Heterogeneous Catalysts For The Transesterification Of Soybean Oil Using Alkali Supported On Nanocrystalline Oxides

Eric J. Doskocil and Samuel E. Yoder. University of Missouri - Columbia, W2033 Lafferre Hall, Dept. of Chemical Engineering, Columbia, MO 65211

The reaction of vegetable oil triglycerides to produce biodiesel typically occurs when a triglyceride present in the oil reacts with an alcohol, usually methanol, to form a mixture of alkyl esters that is termed "biodiesel." Current base-catalyzed biodiesel production is performed batch-wise using a homogeneous alkali base catalyst. After reaction, the alkali remains dissolved in the by-product glycerin phase or undergoes saponification with available free-fatty acids. The resulting product mixture is then neutralized with an acid and the resulting alkali salt is discarded. As an alternative, many basic heterogeneous catalysts have been used successfully to produce biodiesel from various vegetable oils at up to 100% selectivity. However, when regenerated, these catalysts lose a significant portion of its biodiesel yield due to large oil molecules that remain trapped in the catalyst pores or leaching of the active catalyst. Not being able to recover full catalytic activity limits their integration into current facilities.

In order to enhance the reusability of heterogeneous catalysts for the production of biodiesel from vegetable oils, we have used high surface area nanocrystalline oxides, primarily MgO and TiO_2, as catalyst supports with minimal pore structure, which reduce the possibility of the large glyceride molecules getting trapped and allow for reusability with higher yield recovery than when using porous supports with similar internal surface area. Attempts to "tether" basic alkali centers on these oxides were performed via the thermal decomposition of supported alkali metal salts and organometallic precursors (such as Li, Na, and K). Catalytic conversions indicate that leaching of the alkali under reaction conditions can be minimized by choosing the proper support and pretreatment conditions. Highly active transesterification catalysts have been produced using Li acetylacetonate that are active for up to seven reaction trials before losing catalytic activity and, most importantly, require no regeneration steps between each reactor charge. The nanocrystalline oxide catalysts present emerging technology to obtain reusable heterogeneous catalysts with high activity for the base-catalyzed transesterification reaction.

160d Kinetics Of Transesterification Reactions Of Lipids

Earl Alley, Rafael Hernandez, Todd French, and Jimmy Cain. Chemical Engineering, Mississippi State University, Box 9595, Mississippi State University, MS 39762

Production of the renewable fuel biodiesel in commercial operations is accomplished by base or acid catalyzed transesterification of acyl glycerides with low molecular weight alcohols. This report is a continuation of our studies of the kinetics of acid catalyzed transesterification in which we are attempting to obtain a better understanding of the complex of reactions and equilibria involved in this process.

In the experiments reported here, we have studied the kinetics of a number of tri-, di-, and monoglycerides in methanol/sulfuric acid and the acid catalyzed equilibria of some of the intermediates in the transesterification reaction of triglycerides. A description of the analytical methodology, and an assessment of the results of relative rates via competition experiments will also be reported.

160g Process for Producing Biodiesel Jatropha Ester Fuel from Lower and Higher Alcohols and Investigations of Its Blending Properties

Asha Immanuel Raju Chaduvula[1], Murali Yugnadhar Nikku[2], Kiran Kumar Nalla[2], Venkata Satya Chappidi[2], and Gangdhar Madabathula[2]. (1) Chemical Engineering, Andhra University, Andhra University, Visakhapatnam - 530 003, AP, INDIA, Visakhapatnam, 530003, India, (2) Chemical Engg Dept, Andhra University, Andhra University, Visakhapatnam - 530 003, AP, INDIA, Visakhapatnam, 530003, India

Much research has been done on biodiesels over the last 20 yr after the oil crisis in 1973. At present, concern about environmental regulations has been the major reason to look for alternative fuel. A significant level in terms of physico-chemical properties of biodiesel has been obtained but there is a lack of full or partial replacement of fossil fuel that needs to be discussed. Different studies have been carried out using jatropha oil as raw material, different alcohols (methanol, ethanol, butanol) as well as different catalysts, homogeneous ones such as sodium hydroxide, potassium hydroxide and sulphuric acid. In this work advantages and disadvantages of technologies are listed. In this the blending properties are also investigated along with fire point, flash point, cetane index, distillation characteristics, kinematic viscosity, sulphur content and carbon residue.

339a The Current Status Of Biodiesel Application Technology

K. Y. Simon Ng[1], Anfeng Wang[1], Haiying Tang[1], Bradley R. Clark[2], and Steven O. Salley[1]. (1) Department of Chemical Engineering and Materials Science, Wayne State University, 5050 Anthony Wayne Dr., Detroit, MI 48202, (2) Alternative Energy Technology, Wayne State University, 5050 Anthony Wayne Dr., Detroit, MI 48202

Biodiesel is a renewable fuel for diesel engines that is derived from natural oils and fats (e.g., vegetable oils, recycled cooking greases or oils and animal fats) and that specifically meets the specifications of ASTM D 6751. It is composed of monoalkyl esters of long-chain fatty acids, produced by the transesterification with alcohol of the above natural oils. Biodiesel is a DOE designated alternative fuel and is registered as a fuel and fuel additive with the Environmental Protection Agency (EPA). Biodiesel offers many benefits over conventional petroleum diesel. It burns cleaner, with net emissions reductions in particulates, hydrocarbons, and carbon monoxide (and with zero to slight increases in NOx). Biodiesel also possesses a high cetane number (averaging over 50) and improves petroleum diesel cetane performance when blended. Since it is naturally low in sulfur content, it also lowers sulfur emissions when blended with petroleum diesel. Biodiesel blending also imparts improved lubricity to petroleum diesel.

Since it is domestically produced, biodiesel shows great potential for reducing U.S. dependence on foreign energy supplies. It provides a “closed economic loop” in that the feedstock can be grown locally, the biodiesel can be produced locally, and the fuel can be used locally. Furthermore, it is evident that very minimal to no infrastructure change is necessary to implement widespread biodiesel use. Biodiesel blends can be used in any diesel engine and can be transported and stored using existing infrastructure. Moreover, pure biodiesel is environmentally non-toxic and biodegradable. With its high energy balance of 3.2 to 1, biodiesel provides a beneficial 78% life cycle CO2 reduction.

In the US, over 80% of the feedstock for biodiesel production is soybean oil. However the biodiesel made from soybean oil is significantly less stable against oxidation than the rapeseed oil-based biodiesel produced in Europe. Oxidation of biodiesel results in increased viscosity, gum formation, increased acidity, increased peroxide value, and all of these changes can be problematic for the fuel delivery system and engine/transmission. Moreover, problems such as filter plugging, corrosion, hardening of rubber components, fusion of moving components, and engine deposits have been observed with low-quality (i.e., out-of-spec) biodiesel.

This presentation will give an overview of the properties of biodiesel, the production process, and the technical barriers facing the application technology of biodiesel in the United States. The effects of metals and antioxidants on the oxidative stability, and the formation of insolubles after low temperature treatment of soy-based biodiesel will be discussed. The formation of insolubles in biodiesel blends may have serious implications to the automotive fuel delivery system in terms of filter plugging and deposits on injectors and critical fuel pump components.

339b Biodiesel Production: An Alternative Fuel

Asha Immanuel Raju Chaduvula[1], Murali Yugandhar Nikku[2], Kiran Kumar Nalla[2], Venkata Satya Chappidi[2], and Gangdhar Madabathula[2]. (1) Chemical Engineering, Andhra University, Andhra University, Visakhapatnam - 530 003, AP, INDIA, Visakhapatnam, 530003, India, (2) Chemical Engg Dept, Andhra University, Andhra University, Visakhapatnam - 530 003, AP, INDIA, Visakhapatnam, 530003, India

The world is confronted with the twin crises of fossil fuel depletion and environmental degradation. The indiscriminate extraction and consumption of fossil fuels have led to a reduction in petroleum reserves. Alternative fuels, energy conservation and management, energy efficiency and environmental protection have become important in recent years. The increasing import bill has necessitated the search for liquid fuels as an alternative to diesel, which is being used in large quantities in transportation, agricultural, industrial, commercial and even domestic sectors. Biodiesel obtained from vegetable oils has been considered as a promising option. The most common catalysts for biodiesel production are homogeneous basic catalysts. In the present paper, a comparison is made of different basic catalysts (sodium methoxide, potassium methoxide, sodium hydroxide and potassium hydroxide) for methanolysis of vegetable oil. All the reactions were carried out under the same experimental conditions in a batch stirred reactor and the subsequent separation in a separating funnel.

339c Effect Of Major And Minor Components Of Biodiesel On The Lubricity Of Petroleum And Synthetic Fuel

Kapila Wadumesthrige, Anfeng Wang, Steven O. Salley, and K. Y. Simon Ng. Department of Chemical Engineering and Materials Science, Wayne State University, 5050 Anthony Wayne Dr., Detroit, MI 48202

Increasingly strict regulations on the sulfur content of commercial petroleum diesel fuels as required by the United States, Europe, and elsewhere, result in a decrease in the lubricity of these fuels. This has led to the failure of engine parts such as fuel injectors and pumps, because they are lubricated by the fuel itself. The poor lubricity of ultra low sulfur diesel (ULSD) requires additives or blending with another fuel of sufficient lubricity to regain lubricity. Previous studies indicate that low blend levels (1-2 %) of biodiesel derived from vegetable oil can restore lubricity to (ultra-)low-sulfur petroleum-derived diesel. In this work, the lubricity effects of major components of biodiesel (fatty acid methyl esters) and minor components such as antioxidants, phospholipids, Sterols, free fatty acids and glycerol compounds on ULSD, and synthetic fuel, S-8 were studied using High Frequency Reciprocating Rig (HFRR) and Scuffing Load Ball On Cylinder Lubricity Evaluator (SLBOCLE) test methods. Systematic study using distilled biodiesels of various feedstocks reveled that blending distilled biodiesel cannot increase the lubricity of ULSD and S-8 as compared to blending with undistilled biodiesel. Adding minor components present in biodiesel especially polar phospholipids can increase the lubricity of ULSD and S-8 considerably. The effects of individual biodiesel components with various chemical and physical properties on diesel lubricity were also examined and used to develop an empirical mathematical formula to predict biodiesel lubricity. HFRR lubricity measurements were performed at 60 °C according to ASTM D-6079 test method and SLBOCLE measurements were performed at 25 °C and 50% relative humidity as described in ASTM D-6078 test method. The severely hydrotreated ULSD has a HFRR mean wear scar diameter (WSD) of 430 μm and a SLBOCLE maximum load of 3200 g; while the synthetic fuel, S-8, has a HFRR mean wear scar diameter of 679 μm which is much higher than the Diesel Fuel oil specification as well the requirement for Engine Manufactures Association (EMA). Blending both ULSD and S-8 with up to 2% v biodiesel shows a sharp decrease of WSD. The WSD levels off at around 200 μm regardless of the type of fuel or the type of blending biodiesel. This constant 200 μm value may be due to the formation of a uniform lubricating layer at metal-metal contact interface. Further addition of biodiesel did not show any further lubricity improvement In order to elucidate the origin of biodiesel lubricity, several major and minor components present in biodiesel were added to ULSD and S-8. With 2 % v of the Methyl ester of linoleic acid, the lubricity of ULSD was increased only by 12% (calculated based on % wear scar diameter reduction); whereas 2% of biodiesel can improve ULSD lubricity by about 50%. Interestingly, vacuum distilled biodiesel with no minor components such as sterols, antioxidants and phospholipids only improved the lubricity of ULSD by 28%. The lubricity enhancement mechanism of these minor components will be discussed. These results will help to develop better and reliable lubricity improves for commercial applications.

339d Biodiesel From Sewage Sludge

Kaiwen Liang, Rafael Hernandez, Hossein Toghiani, and Todd French. Chemical Engineering, Mississippi State University, 330 Swalm Circle President Dr., Mississippi State University, MS 39762

Primary and activated sludges generated in wastewater treatment facilities consist of a variety organic matter and microorganisms. The lipids in this sludge could be used to generate biodiesel, a mixture of fatty acid methyl esters that conforms to the ASTM D6751 standard. Biodiesel is sold commercially as a blend with petroleum diesel. In 2006 biodiesel production was approximately 80 million gallons. In this work, primary and secondary sludges were collected from a municipal wastewater treatment plant located in Tuscaloosa, AL. After collection, sludges were concentrated by centrifugation, and dried using a freeze dryer. Sludges were also homogenized prior to each experiment. Lipids in dry sludge (~5% moisture) were converted to biodiesel by in-situ transesterification. In-situ transesterification combines extraction and transesterification into one step. Gas chromatography was used to analyze the fatty acid methyl esters in the sample. The biodiesel generated from the sludges was also tested for ASTM compliance. ASTM testing included viscometry, flash point, Cu-corrossion test, cloud point, C resident, distillation, and S analysis. It is estimated that wastewater treatment facilities could provide approximately 800 million gallons of feedstock for biodiesel production. The presentation will include an economic analysis of the generation of biodiesel from sewage sludge.

339e In-*situ* Transesterification Of Chinese Tallow Oil To Biodiesel

Jaricus A. Whitlock and Dr. Rafael Hernandez. Dave C. Swalm School of Chemical Engineering, Mississippi State University, Box 9595, Mississippi State, MS 39762

Biodiesel is a renewable fuel produced from vegetable oils and animal fats that conforms to the ASTM D6751 standard. In 2006, biodiesel production was approximately 80 million gallons. If the industry continues to grow at the present rate, this biofuel could contribute to a significant reduction of the US dependency on foreign oil. To maintain this rate of growth, other oil feedstocks will have to become available. Several new and innovative feedstocks have captured the interest of biodiesel producers, one of those being Chinese tallow. The Chinese tallow tree (sapium sebiferum) seeds produced contain approximately 40% - 45% extractable material. The oil yield per acre is 10 times higher than soy bean oil.

Conventionally, Chinese tallow oil would be extracted from the seed prior to transesterification using hexane or a press. However, transesterification of the oil through traditional methods has proven to be difficult. Mostly composed of saturated fatty acids, Chinese tallow oil solidifies when exposed to ambient temperatures. Excess heating is required to maintain the oil's liquid phase so that transesterification can be successfully performed. In-*situ* transesterification of Chinese tallow seeds simultaneously combines the steps of lipid (oil) extraction and transesterification to produce biodiesel. Integration of these stages could reduce the need for excess heating while maintaining a high conversion to biodiesel. The paper will describe results obtained on the in-*situ* transesterification of Chinese tallow seeds, biodiesel ASTM testing, and processing economics.

Intermediate Temperature Conversion of Bio-Oil to Synthesis Gas for Distributed Hydrogen Production

Jonathan R. Marda[1], Anthony M. Dean[1], Stefan Czernik[2], Robert J. Evans[2], Richard J. French[2], and Matthew A. Ratcliff[2]

[1]Colorado School of Mines, Golden, CO
[2]National Renewable Energy Laboratory, Golden, CO

Introduction

With the world's energy demands rapidly increasing, it is necessary to look to sources other than fossil fuels, preferably those that minimize greenhouse emissions. One such renewable source of energy is biomass, which has the added advantage of being a near-term source of hydrogen. Over 500 million tons of biomass could be available in the U.S. at less than $50/ton, which can be converted to 50 million tons of hydrogen.[1] The challenges for the variety of feedstocks that will be used include handling and drying, regional and seasonal availability and variability, and the impurities that could be generated and have an impact on conversion technology and hydrogen purity. Processes must be feedstock flexible, and minimize costs for feedstock collection, transport, and processing.

While there are three main potential routes to produce hydrogen from biomass thermally, given the near-term technical barriers to hydrogen storage and delivery, technologies that can be distributed such that hydrogen is produced at or near the point of use are attractive. The first potential route involves gasification of biomass to synthesis gas ($CO + H_2$) at high temperatures in which air, oxygen, or steam is used as oxidizers. In the second case, bio-oil is first produced via fast pyrolysis at relatively moderate temperatures and then reformed to hydrogen and carbon dioxide. The third route includes a variety of high-pressure processes known as wet gasification or aqueous-phase reforming that is especially suitable for high-moisture content biomass. Only the second method allows for the processing of biomass and the production of hydrogen at different locations utilizing low-cost biomass resources and existing infrastructure for hydrogen distribution while saving on high-cost transportation of biomass and hydrogen. This approach is especially well suited for smaller-scale reforming plants located at hydrogen distribution sites such as filling stations. Additionally, there is the potential for automated operation of the conversion system.[2]

The thermally reactive compounds (e.g., anhydro-sugars, phenolics) found in bio-oil do not evaporate cleanly. They tend to decompose and react, with the potential of forming carbonaceous deposits or being converted to aromatic hydrocarbons, which are thermally stable and more difficult to convert to hydrogen. Because of this, conventional fixed bed reformers have not proven efficient for this highly reactive feedstock. Reactors that fluidize or circulate the catalyst are much more suited for this application, but are not an optimal choice for small-scale and unattended operation.

This research addresses the challenge of distributed hydrogen production with the target of producing hydrogen at $3.80/kg by 2012.[3] A system has been developed for volatilizing bio-oil with manageable carbon deposits using ultrasonic atomization and by modifying bio-oil properties, such as viscosity, by blending or reacting bio-oil with methanol.

Non-catalytic partial oxidation of bio-oil is then used to achieve significant conversion to CO with minimal aromatic hydrocarbon formation by keeping the temperature at 650°C or less and oxygen levels low. The non-catalytic reactions occur primarily in the gas phase. However, some nonvolatile components of bio-oil present as aerosols may react heterogeneously. The product gas is passed over a packed bed of precious metal catalyst where further reforming as well as water gas shift reactions are accomplished completing the conversion to hydrogen.[4]

The approach described above requires significantly lower catalyst loadings than conventional catalytic steam reforming due to the significant conversion in the non-catalytic step. The goal is to reform and selectively oxidize the bio-oil and catalyze the water gas shift reaction without catalyzing methanation or oxidation of CO and H_2, thus attaining equilibrium levels of H_2, CO, H_2O, and CO_2 at the exit of the catalyst bed.

Model compounds are used to establish the underlying chemical understanding of the process, as well as identify optimized conditions. To date partial oxidation and pyrolysis experiments using neat methanol (650°C) have been performed and compared to predictions of a detailed kinetic model. Kinetic modeling is useful in verifying that the chemistry occurring in the system is consistent with what is expected to occur in gas phase chemistry. The model used is a rule-based mechanism that has been developed to predict both oxidation and pyrolysis of hydrocarbons and has been extended to methanol. The premise behind the model is that three types of free radical reactions (dissociation/recombination, hydrogen abstraction, and b–scission/radical addition) dominate the kinetics.

Experimental

Bio-oil (mixed with varied amounts of methanol to reduce the viscosity and homogenize the bio-oil) or selected bio-oil components (such as methanol) are introduced at a measured flow rate through the top of a vertical reactor (see Figure 1) using an ultrasonic nozzle. The nozzle creates a fine mist (at ambient conditions), which allows the bio-oil to flow down the center of the reactor at a low, steady flow rate. Additionally, the fine mist allows for intimate mixing with oxygen and efficient heat transfer, providing optimal conditions to achieve high conversion at relatively low temperatures. Generation of the fine mist is especially important for providing good contact between non-volatile bio-oil components and oxygen. Oxygen and helium are also delivered at the top of the reactor via mass flow meters with the amount of oxygen being varied to maximize the yields of H_2 and CO and the amount of helium being adjusted such that the gas phase residence time in the hot zone is ~0.3 and ~0.45 s for bio-oil and methanol experiments, respectively.

The reactants are heated to 650°C using a five-zone furnace. A catalyst bed is located at the bottom of the reactor tube. To date, catalyst screening experiments have used Engelhard noble metal catalysts. The catalysts used for these experiments were 0.5 % rhodium, ruthenium, platinum, and palladium (all supported on alumina). Experiments were performed using pure alumina as well. Both the catalyst type and the effect of oxygen on the residual hydrocarbons and accumulated carbon containing particulates were investigated. Non-catalytic experiments (primarily homogeneous cracking) uses a bed of quartz particles placed to capture any deposits that are formed in the volatilization and cracking zones.

The inner reactor effluent is quenched by a flow of 10 SLPM He which serves to sweep the products quickly (~0.03 s) to a triple quadrupole Molecular Beam Mass Spectrometer (MBMS) for analysis. The MBMS serves as a universal detector and allows for real time data collection. The study of pyrolysis by MBMS has been described previously.[5] The gases exiting the outer tube are expanded through an orifice on the apex of a sampling cone into the first vacuum stage (40 mtorr). The free-jet expansion quenches the products and allows light gases, high-molecular-weight compounds, and reactive products, to be sampled. A molecular beam, collimated through a second expansion, enters an ion source, where 25 to 50 eV electron impact ionization is used to form ions that are analyzed by a quadrupole mass filter. The dilution of the reactor effluent reduces the potential problems caused by matrix effects associated with the MBMS analysis. Argon is used as an internal standard in the quantitative analysis of all the major products (CO, CO_2, H_2, H_2O, and benzene) as well as any residual carbon, which is determined by subsequent oxidation of carbon (monitored as CO_2) after shutting off the feed and maintaining the oxygen/helium flow. Any carbon left on the walls of the reactor (measured by weighing) is also counted as residual carbon.

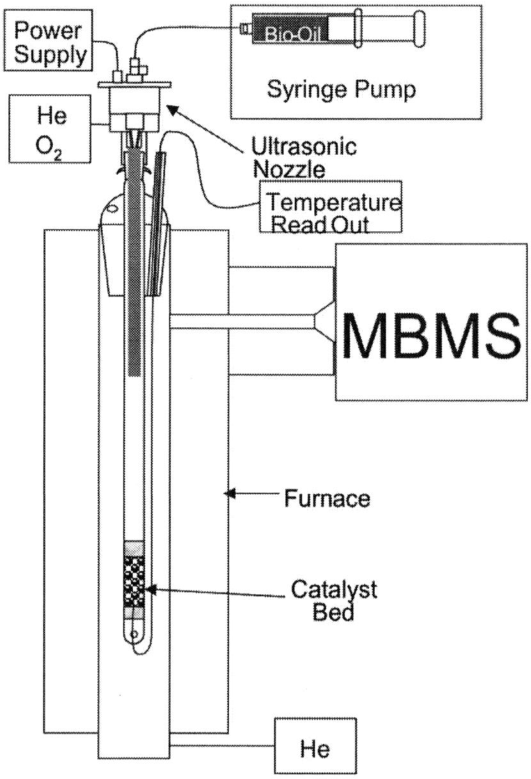

Figure 1. Experimental apparatus for converting bio-oil to hydrogen.

Results and Discussion

Non-Catalytic Partial Oxidation

Performance of the bio-oil injections system with respect to non-catalytic conversion (primarily in the gas phase) was tested with a 50:50 (wt. %) bio-oil:methanol solution. These experiments were performed to determine the extent of low temperature oxidation of the volatilized bio-oil. Variables were temperature and oxygen concentration. The objective was to break carbon-carbon bonds to attain high yields of CO and H_2 while avoiding oxidation to CO_2 as well as aromatic hydrocarbon formation. The temperature was varied from 500 to 650°C. The oxygen level was varied from that present in the methanol/bio-oil feed to the maximum level that could be added in the gas phase without the onset of significant combustion. During these experiments, it was determined that temperatures above 600°C yielded the most promising results. However, it should be noted that even at temperatures as low as 550°C a significant amount of oxidative cracking occurred without the presence of a catalyst.

In addition to the experiments using the 50:50 mixture described above, experiments were performed using 70:30 and 90:10 bio-oil:methanol mixtures at 650°C. The results of all experiments performed at 650°C are plotted against an effective O:C ratio where the oxygen due to the water content of the bio-oil is subtracted out. The effective O:C ratios are 0.13, 0.17, and 0.21 less than the actual O:C ratio for the 50:50, 70:30, and 90:10 bio-oil methanol mixtures, respectively.

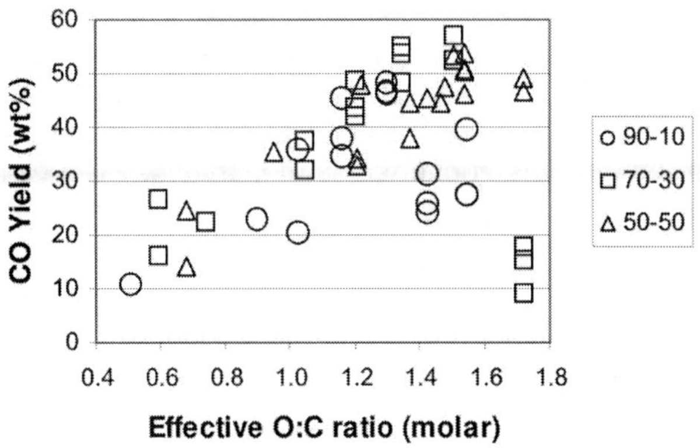

Figure 2. CO yield from bio-oil (excluding estimated CO production from methanol) plotted against effective O:C ratio (oxygen entering as water is removed).

The estimated CO yields from the bio-oil (i.e., after the CO attributed to methanol is subtracted out) are shown in Figure 2; they initially increased as the effective O:C ratio increased. The maximum yields of CO for the 50:50 mixture (~50 %) and the 70:30 mixture (~55 %) were remarkably similar and occurred first at nearly the same effective O:C ratio (~1.5). The maximum (~47 %) for the 90:10 mixture occurs at a lower O:C ratio (~1.3). For both the 70:30 and 90:10 mixtures, the CO yield appeared to decrease significantly at O:C ratios higher than where the maximum occurred, whereas CO yield for the 50:50 mixture did not decrease.

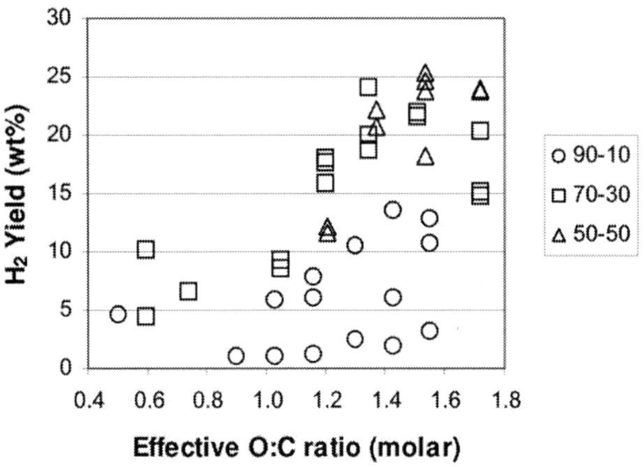

Figure 3. Overall H_2 yield (including CO production from methanol) plotted against effective O:C ratio (oxygen entering as water is removed).

The overall hydrogen yields are shown in Figure 3. Just as with the maximum CO yields, the maximum H_2 yield for the 50:50 (~23 %) and 70:30 (~22 %) mixtures are nearly identical and occur at an effective O:C ratio of about 1.5. The maximum (~15 %) for the 90:10 mixture occurs at a lower O:C ratio (~1.4). It should be noted that the maximum H_2 and CO yields occurred at nearly the same O:C ratios for the 70:30 and the 90:10 mixtures.

Table 1 shows the yields of all products that were quantified at the maximum CO Yield for each mixture. While the CO_2 and H_2O yields are similar for all mixtures, the 50:50 and 70:30 mixtures have much higher hydrogen yields. The 70:30 and 90:10 mixtures have much lower benzene yields. The 90:10 mixture has the highest residual carbon yield reflecting the fact that more carbon was deposited on the walls of the reactor near the entrance due to poor nozzle performance. Here, residual carbon is defined as carbon burned from the reactor after the bio-oil feed is turned off plus any carbon left on the walls of the reactor. The majority of the residual carbon in all cases is due to carbon left on the reactor walls.

Table 1. Product distribution of bio-oil experiments at maximum CO yield. The values shown are the average yields for the conditions that yielded the maximum CO. For carbon containing species, the yields are amount of carbon entering as bio-oil that is converted to that species on a weight percent basis. For H_2 and H_2O, the yields are the amount of hydrogen entering the system (either as bio-oil or methanol) that is converted to that species on a weight percent basis. The error shown is the standard deviation of two to four runs.

Mixture	CO_{max} ($O:C_{effective}$)	$(CO_2)_{CO_{max}}$	$(\text{Benzene})_{CO_{max}}$	$(\text{Residual C})_{CO_{max}}$	$(H_2)_{CO_{max}}$	$(H_2O)_{CO_{max}}$
50:50	50 ± 3 (1.5)	9 ± 2	3 ± 2	3 ± 2	23 ± 3	47 ± 2
70:30	55 ± 2 (1.5)	$11.9 \pm .6$	$1.4 \pm .3$	3 ± 2	$21.7 \pm .3$	$55.1 \pm .2$
90:10	47 ± 1 (1.3)	14 ± 2	$0.7 \pm .1$	11 ± 5	8 ± 5	58 ± 9

In summary, the experiments using the 70:30 mixture yielded remarkably similar results to the experiments using the 50:50 mixture. The experiments using the 90:10 mixture were quite different. This difference may have been due to decreased performance from the ultrasonic nozzle when using the more viscous liquid. We have recently acquired an ultrasonic nozzle that has the capacity to handle more viscous liquids, which should allow for better results.

Catalytic Auto-Thermal Reforming

The products from oxidative cracking will include some residual oxygenates and low molecular weight hydrocarbons, such as benzene. The oxidation and reforming of these followed by the water gas shift reactions will be accomplished by low temperature auto-thermal catalysis. Catalyst testing experiments at 650°C using the 50:50 bio-oil:methanol mixture have been initiated using commercially available catalysts (Engelhard). A comparison of packed beds of 0.5% rhodium, palladium, platinum, and ruthenium catalyst on alumina was performed as a benchmark to establish the competitive reaction pathways for syngas reactions, such as oxidation, water-gas shift, and steam reforming. It should be noted that the catalysts used were chosen because they were on hand. The catalysts are not considered optimal in size or loading and results presented are preliminary.

The results are shown in Table 2. The equilibrium values and gas phase only (non-catalytic) results are shown for comparison. The concentration of oxygen and water for the gas phase reaction were held constant for the catalyst experiments, which means that only steam was available as a reactant gas on the catalyst since all oxygen is used up in the gas phase section.

Table 2. Comparison of catalysts with equilibrium and gas phase product distributions at an O:C ratio of 1.3. Values expressed as % of carbon or hydrogen.

	CH3OH	CO	CO2	H2	H2O
	% conversion	% of C	% of C	% of H	% of H
Equil.	100	73	23	81	12
Gas	49	39	7	13	38
Alumina	39	36	7	9	38
Pd	82	51	10	34	41
Pt	88	45	12	31	45
Rh	98	56	25	71	19
Ru	75	48	10	30	43

The reaction with residual hydrocarbons on the catalyst was complete although conversion to carbon is a likely reaction pathway. As with the gas phase studies, the solid carbon was quantified when the feed was turned off and the oxygen becomes available for carbon burn off. To date, no attempt has been made to characterize the carbon that accumulates on the catalyst. The Rh catalyst was the closest to equilibrium conditions. Although benzene yields are not shown in the table, it should be noted that they were similar to those seen in the gas phase indicating that these particular catalysts were not effective for converting benzene with the limited amount of steam present. The challenge for the integrated system will be to selectively reform the benzene and other residual hydrocarbons while catalyzing the water gas shift reaction without catalyzing methanation and CO oxidation. The goal will be to attain equilibrium levels of H_2, CO, H_2O, and CO_2 at the exit of the catalyst bed while maintaining steady-state conditions, which means adequate carbon conversion in the first phase of the catalyst bed.

Kinetic Modeling of Methanol

An existing kinetic model was extended to predict the results of the methanol experiments discussed in our previous quarterly report. The current model (CHEMKIN format) is a rule-based model that has been used to predict both oxidation and pyrolysis of hydrocarbons. The premise behind the model is that three types of reactions can be used to describe the system: dissociation/recombination, hydrogen abstraction, and β–scission. Without making any changes to the model, simulations were performed to determine the agreement between the data and the model. A model developed by Held and Dryer[6] was also used in simulations as a means of comparison. It should be noted that during simulations, temperature changes due to reaction were neglected. From these simulations we determined that adjustments to our model were required in order to better predict the data. Using both rate of production analysis and sensitivity analysis, key reactions were identified.

The current model predicts methanol conversion better at lower O:C ratios, but the Held-Dryer model does better at higher O:C ratios. A potential reason that the current model under predicts the conversion at higher O:C ratios is that there is more potential for a large temperature change (as much as 20°C higher based on recent experiments) due to more oxidation occurring. If this explains the under prediction in our model, it would mean that the Held-Dryer model is running too fast. There are large discrepancies in the prediction of both CO and H_2 by both models, which are due in large part to the discrepancies in the conversion of methanol. The current model over predicts formaldehyde (CH_2O) production for all O:C. However, the prediction of CH_2O is much improved. The Held-Dryer model does a much better job of predicting CH_2O. Prediction of CO_2 is in need of improvement. The current model over predicts CO_2, while the Held-Dryer model under predicts it. Both models do an excellent job of predicting H_2O.

Although the model is in need of improvement, the rough agreement between experimental results and model predictions is sufficient to believe that the reaction of methanol is occurring in the gas phase.

Conclusions

These results show that a low temperature oxidative cracking stage will reduce the required catalyst load in the reforming of bio-oil and that a clean syngas can be produced without the use of high gasification severity conditions, namely temperatures greater than 650°C and excessive amounts of water to drive the carbon reforming reactions to completion. Ongoing developments in the project are examining the parametric effects of variation in process variables (temperature, oxygen, and steam), the removal of CO_2 from the system by adsorption within the catalyst bed, and the impact of variations in the composition of the bio-oil. Model compounds are being studied to develop a mechanism of homogeneous oxidative cracking, which will add insight to the chemistry of the gas phase reactions. Catalysis studies will be expanded to determine optimum materials that will minimize cost and minimize the formation of aromatics and other syngas contaminants.

Acknowledgement

The authors gratefully acknowledge the financial support from the Hydrogen Fuel Cell and Infrastructure Technology Program of the US DOE.

References

1. Milbrandt A., A Geographic Perspective on the Current Biomass Resource Availability in the United States, *National Renewable Energy Laboratory: Golden, CO, TP-560-39181*, 2005.
2. Czernik, S., Elam, C., Evans R., and Milne, T., Thermochemical routes to hydrogen from biomass – a review, *In Science in Thermal and Chemical Biomass Conversion*, Bridgwater AV, Boocock DGB, eds., CPL Press: Newbury, UK, pp.1752–1761, 2006.
3. U.S. Department of Energy.,*Hydrogen, Fuel Cells and Infrastructure Technologies Program, Multi-Year Research, Development and Demonstration Plan, Section 3.1 Hydrogen Production*, U.S. Department of Energy, Office of Energy Efficiency and Renewable Energy: Washington, D.C, 2006.
4. Evans, R.J., Czernik, S., French, R., and Marda, J., Distributed Bio-Oil Reforming, *DOE Hydrogen Program FY2006 Annual Progress Report*, 2006.
5. Evans, R. J., and Milne, T. A., Molecular characterization of the pyrolysis of biomass, *Energy & Fuels*, Volume1, pp. 123-137, 1987.
6. Held, T.J., and Dryer, F.L., A comprehensive mechanism for methanol oxidation, *International Journal of Chemical Kinetics*, Volume 30, pp. 805-830, 1998

339g "Karanji Oil-Diesel Blend": As An Alternative Fuel for Diesel Engine

Asha Immanuel Raju Chaduvula[1], Murali Yugandhar Nikku[2], Kiran Kumar Nalla[2], Venkata Satya Chappidi[2], and Gangdhar Madabathula[2]. (1) Chemical Engineering, Andhra University, Andhra University, Visakhapatnam - 530 003, AP, INDIA, Visakhapatnam, 530003, India, (2) Chemical Engg Dept, Andhra University, Andhra University, Visakhapatnam - 530 003, AP, INDIA, Visakhapatnam, 530003, India

With depleting oil resources and negative environmental impacts associated with use of biofuels, there is renewed interest in biomass based fuels. Now, edible oil like Pongamia, jatropha, algemone, mahua, castor etc can be used for the production of biodiesel. Pongamia pinata is multi purpose plant with many attributes and enormous potential for biodiesel production. Given the ballooning cost of importing crude oil, biodiesel from pongamia pinata makes eminent sense as supplement as it not only burns cleanly than petroleum diesel, it does not require any modification in the present ignition engines for diesel or even to the current distribution infrastructure of storage tank and petrol stations that deliver fuel diesel to the customer. Karanji oil diesel blend is an alternate fuel for diesel engine.

Catalytic Effect of Group IV Transition Metals on Oxidative Stability of Biodiesel

Bradley R Clark, Anfeng Wang, Steven O. Salley, K. Y. Simon Ng

Alternative Energy Technology,

Department of Chemical Engineering and Materials Science,

Wayne State University, Detroit, MI. 48202

EXTENDED ABSTRACT

Biodiesel is a major source of liquid biofuels. Although the consumption of biodiesel has experience strong growth in recent years, fuel quality problems have been reported. Since biodiesel is used in compression ignition engines, fuel injector fouling is especially problematic. Thus, the formation of high molecular weight species as a result of biodiesel oxidation is a significant problem. Factors that increase the rate of oxidation include light, heat and metals. The effect of group IV transition metal ion catalysis of biodiesel oxidation is investigated here.

There are reported interactions between metal ions and triglycerides, fatty acids or fatty acid methyl esters (biodiesel). Metal ions are catalytic for the initiation and propagation stages of free radical oxidation. In triglycerides and fatty acids, the catalytic detrimental effects of copper and iron have been described. The reported effects are an increase of peroxide value and decrease of induction period. The catalytic effects of cobalt, manganese and nickel are also reported. In biodiesel, the catalytic effects of copper and iron are reported frequently. The effects include an increase in carboxylic acid formation and a reduction of induction period. The ability of cerium to reduce induction period has also been reported. Since biodiesel is most often used in blends, catalytic metals could be introduced from petroleum diesel. Transition metals at the ppm levels in petroleum diesel from Asia have been reported.

Based on a brief literature review, it is clear that the catalytic effects of transition metal ions on the oxidative stability of biodiesel have not been fully elucidated. While these reports indicate the catalytic effects of different metals, the choice of metals was sporadic and the concentration ranges are limited. The group IV transition metal ions selected for investigation are: V, Cr, Mn, Fe, Co Ni, Cu and Zn. The catalytic effects of Mn, Fe, Co, Ni and Cu ions have been previously reported although there was little or no attempt to determine the effect of concentration on the rate of oxidation. Zn was added because the catalytic effects of the metal have been reported and because it is present in brass fittings. Cr was added because it is present in stainless steel. V was added because of its presence in crude petroleum. The metals were added as nitrate or chloride salts pre-dissolved in anhydrous methanol. Since the rate of oxidation is influenced by the presence of naturally occurring antioxidants, different feedstocks: soybean oil and cottonseed oil biodiesel were tested.

The objectives of this research are to determine: 1) which of these metals are catalytic, 2) what is the relationship between metal ion concentration and induction period change and 3) what is the threshold concentration of each catalytic metal such that the reduction of induction period was 20%? The 20% value was determined empirically based on the experience of the researchers.

The results of oxidative stability measurement are reported as Δ% induction period. This equation was employed as a way to compensate for variations in the operation of the instrumentation. Replicates of a given sample of biodiesel do not result in a single induction period. Rather, it produces a range of values. It was decided that this equation would produce more consistent results given the unavoidable variations in the data.

First we discuss high activity metals. "High Activity" is defined as the capacity to produce a 20% reduction of induction period at a concentration level less than 2 ppm. The two most active metals for B100 from soybean oil are Cu and Cr. The effects of Cu have been previously reported. There is a "leveling off" effect in the concentration range of 1 to 2 ppm. This is expected because the Δ% IP equation cannot generate values lower than -100. The "leveling off" effect has an intuitive interpretation since catalysis effects would be limited by the kinetics of the rate determining reaction. The most active metals for cottonseed oil are Cr, Cu, Co and V. Cu and Cr are highly active in both feedstock but cottonseed is more sensitive to the effects of Co and V.

Next we discuss medium activity metals. "Medium Activity" is defined as the capacity to produce a 20% reduction of induction period at a concentration level less than 10 ppm. Co and V, which were high activity metals in cottonseed B100, are now medium activity metals in soybean oil. Although the concentration range is higher, both show the ability to reduce IP by more than 80%. Mn and Zn are medium activity metals in both feedstocks.

Next we discuss low activity metals. “Low Activity” is defined as the capacity to produce a 20% reduction of induction period at a concentration level less than 100 ppm. Fe and Ni are low activity metals in both feedstocks. In both cases Fe shows a greater capacity to decrease induction period than Ni.

The Rancimat induction period is essentially a measure of the ability of a sample of biodiesel to resist free radical oxidation. The presence of naturally occurring antioxidants will increase the induction period. Antioxidants that are found in triglycerides are likely to carry through to the biodiesel. These can be separated into two broad categories: tocopherols and isoflavones. Tocopherols are readily oxidized and would not be expected to offer significant protection during the Rancimat test. It is hypothesized that the difference between the Rancimat results of soybean oil and cottonseed oil can be attributed to the differences in naturally occurring isoflavones. There is precedence for this interpretation in the literature. Quercetin and kaempferol are typical isoflavones found in cottonseed. Daidzin and genistein are typical isoflavones found in soybeans. Isoflavone may function to inhibit free radical oxidation in two ways. They may function as peroxy-radical scavengers and thus terminate the sequence of free radical oxidation. They may also function a chelators, binding transition metals into stable complexes thereby inhibiting catalysis.

The concentration of different isoflavones present in biodiesels of different feedstocks is an open question. Variations in triglycerides will inevitably produce variations in the biodiesel manufactured from them. Variations in the method of manufacture must also be considered. Any attempt to make broad generalizations about isoflavone concentrations in currently produced biodiesel is problematic. Yet, the capacity of different isoflavones to act as metal chelators and free radical chain terminators is a good working hypothesis for the differences observed in preceding data.

The solubility of metal ions during the Rancimat test is another issue to consider. All the salts listed were soluble in anhydrous methanol. But that does not guarantee that they will be soluble in biodiesel. Since solubility is a function of concentration, the metals with the lowest catalytic activity will be most affected. The question is further complicated by the continuous accumulation of oxidation products which may also change metal ion solubility.

In both feedstocks, the Rancimat tests of Fe and Ni produced significant amount of colored insoluble material inside the reaction vessel. At the end of the tests, brown and light green precipitates were observed inside the reaction vessel for Fe and Ni tests, respectively. Under the conditions of the Rancimat test these precipitates could be oxides, hydroxides, organic acid salts or some combination of these. The limited solubility of these ions would remove them from solution and diminish their ability to act as catalysts although it is still possible that these metal precipitates would function as a catalytic surface. The results for Zn are also influenced by solubility. A small amount of grey precipitates were observed inside the reaction vessels at the highest Zn ion concentrations.

Biodiesel, as any other liquid transport fuel, needs specifications to ensure quality. From the results presented above, it is clear that limits on catalytic metals must be established to limit free radical oxidation leading to fuel quality degradation. This necessary work has already begun with the release of ASTM Designation: D 6751-07. While this specification covers calcium, magnesium, sodium and potassium it does not cover any transition metals. The present work is offered as guidance to those who want to include limits on transition metals in future specifications. After discussions with different interested parties at the National Biofuels Energy Lab and the National Biodiesel Board it was determined to set the limit at a metal concentration sufficient to cause a -20% change of induction period. This value is a starting point and may need to change as further research findings are revealed.

In many ways the conditions of the Rancimat induction period test are not typical of normal biodiesel storage. The high temperatures and continuous air stream found in this test would certainly be avoided in normal storage. Yet, this does not imply that the Rancimat test is unrealistic. It is possible that these conditions may approximate the conditions of fuel recirculation in a modern diesel engine. The literature indicates that biodiesel oxidation proceeds by the usual free radical mechanism of initiation, propagation and termination of hydroperoxides formed at bis-allylic sites. The results of autoxidation under the conditions of the Rancimat test are a complex mixture. Oxidation products of both low and high molecular weight are to be expected.

The determination of metal concentration limits must be a function of the induction period data. These concentration values will be interpolated from equations derived from the data. Such equations must conform to the kinetics of the metal catalyzed oxidation reactions described in the literature. Thus, it is important to review the mechanisms of metal catalysis of free radical oxidation. Transition metals can catalyze reactions in both the initiation and termination stages.

The classic "Lipid Oxidation" by Frankel provides the kinetics analysis needed. Our initial kinetics analysis will focus on the metal catalysis of the propagation step of free radical oxidation. A kinetics analysis of metal catalysis of initiation and termination is also discussed. The termination of the free radical chain would proceed by a variety of reactions. For the purpose of this analysis we assume that the peroxyl termination reaction predominates. With the reaction mechanism and assumptions stated here, there is a functional relationship between the added metal concentration and the inverse of the square of the induction period. The functional relationship for antioxidant termination mechanism is between the added metal concentration and the inverse of the induction period. The case of metal catalyzed termination is more complex than the previous two. The functional relationship associated with the added metal concentration is the γth root of the inverse of the induction period where γ is some negative number.

Of the models devised there are three choices of termination mechanisms leading to three different functional relationships between metal concentration and induction period. Of these, the metal catalyzed termination mechanism leads to the most complex model. This termination mechanism is further complicated by the existence of metal-hydroperoxide complexes. Such complexes are described in the literature and are reported to cause non-linear effects in the rate of hydroperoxide decomposition. The argument for the effects of hydroperoxide complexes is as follows. If hydroperoxide concentration exceeds metal concentration then all metal ions will be bound in hydroperoxide complexes. The result is hydroperoxide decomposition and peroxyl initiation. If metal ion concentration exceeds hydroperoxide concentration then some metal ions are "free" to participate in the termination mechanism, thus slowing the rate of oxidation. A similar argument can be made for alkyl initiation. Instead of participating in the termination mechanism, the "free" metal ions would participate in another initiation mechanism thus accelerating the rate of oxidation. It is difficult to predict which mechanisms will predominate, but the polarity of the solvents has been described as a major influence. In any case, it is reasonable to conclude that non-linear effects will increase as metal ion concentration increases. The peroxyl radical termination mechanism has the strongest basis in the literature. The equation derived from this mechanism will be used to determine the -20% induction period limits.

At the beginning of the Rancimat test, hydroperoxide concentrations are at a minimum and antioxidant concentrations are at a maximum. As the test proceeds, the hydroperoxide concentration increases and the antioxidant concentration decreases.

When the induction period time is reached, antioxidant concentration has decreased to a point where it is not longer able to "damp" the rate of oxidation and hydroperoxide values have reached a constant. This is the inflection point of the Rancimat curve where the rate of oxidation increases dramatically.

Group IV transition metals are catalysts for the oxidation of biodiesel. Catalytic effects are dependent of the choice of metal. Copper and chromium exhibit the greatest catalytic activity. Iron and nickel exhibit the least catalytic activity. Vanadium, manganese, cobalt and zinc have catalytic activities greater than iron or nickel. An analysis of the data based on the reported mechanisms of free radical termination and the corresponding kinetic equations was performed. At low metal concentrations, the formation of metal-hydroperoxide complexes leads to hydroperoxide decomposition. Thus the peroxyl termination mechanism predominates and the metal concentration is a linear function of the inverse of the square of the induction period as predicted by the kinetics. This equation allows the calculation limit concentrations sufficient to cause a -20% change of induction period. At high metal concentrations nonlinear effects are observed. These nonlinear effects have been previously reported and are the result metal ions participating in termination and/or initiation mechanisms other than hydroperoxide decomposition.

394b Optimization And Reaction Kinetics Of The Production Of Biodiesel From Castor Oil

Scott D. Crymble[1], Dr. Rafael Hernandez[1], Todd French[2], and Earl Alley[2]. (1) Dave C. Swalm School of Chemical Engineering, Mississippi State University, P.O. Box 9595, 330 Swalm Chemical Engineering Building, Mississippi State, MS 39762, (2) Chemical Engineering, Mississippi State University, Box 9595, Mississippi State, MS 39762

Ricinus communis, commonly known as the castor plant, has been cultivated for millennia as a source of oil for lamps and other applications. The seeds of the castor plant contain 35% to 55% oil. The major component of castor oil is ricinoleic acid, which comprises 90% of the fatty acid profile. Ricinoleic acid is a monounsaturated, 18-carbon fatty acid with a hydroxyl functional group at C-12. This hydroxyl group causes castor oil to be highly polar, simplifying the mixing of the oil with methanol for the biodiesel reaction. Castor oil is more viscous than typical oils, which poses problems in the resulting biodiesel.

A study of the esterification reaction kinetics of converting the castor oil into biodiesel was performed to determine reaction kinetics. This information can be used to engineer a biodiesel production facility with castor oil as a feedstock. Factors considered in the kinetics study included reaction temperature, catalyst concentration, ratio of methanol to oil, and reaction time. The resulting biodiesel and blends of castor biodiesel with other biodiesels (e.g., soy bean oil) were tested for compliance with appropriate ASTM standards.

Production of biodiesel from castor oil would provide two benefits: (1) a source of oil that does not compete with the food supply, and (2) reduce the US dependency on foreign oil by providing a domestic alternative to petroleum diesel.

394c The Possibility of Producing Biodiesel Fuel from Wood Pulping Byproducts

Yan Pu[1], Paul D. Fleming III[2], Harold Hladky Sr.[1], and Dewei Qi[3]. (1) Chemical Engineering, Paper Engineering and Imaging, Western Michigan University, 1504, 1A, Concord Place Dri., Kalamazoo, MI 49009, (2) Department of Paper Engineering, Chemical Engineering and Imaging, Western Michigan University, Kalamazoo, MI 49008, (3) Paper Engineering, Chemical Engineering & Imaging, Western MichiganUniversity, Parkview Campus A227,PCI, Kalamazoo, MI 49008

As energy demand increases and the fossil fuels are limited and un-renewable, research has been directed to find an alternative and renewable fuel. Biodiesel fuels produced from biomass material is a renewable and promising substitute fuel for petroleum diesel fuel, and it is relatively environmental friendly. A lot of research has been carried out dealing with biodiesel fuel production, which is generally made from vegetable oil or animal fats. The Pulping process involves a huge amount of lignocellulosic materials, which are considered to be the largest portion of biomass. There may be a great potential of combining this fact with the biodiesel production, as well as the present biodiesel technologies. This work explores the possibility of preparing biodiesel from kraft pulping byproduct (tall oil fatty acid) and investigates how temperature, pressure, reaction time, and molar ratio affect the reaction.

394d Optimization for the Biodiesel Production Using Response Surface Methodology

Asha Immanuel Raju Chaduvula[1], Murali Yugandhar Nikku[2], Kiran Kumar Nalla[2], Venkata Satya Chappidi[2], and Gangdhar Madabathula[2]. (1) Chemical Engineering, Andhra University, Andhra University, Visakhapatnam - 530 003, AP, INDIA, Visakhapatnam, 530003, India, (2) Chemical Engg Dept, Andhra University, Andhra University, Visakhapatnam - 530 003, AP, INDIA, Visakhapatnam, 530003, India

Response surface methodology(RSM) was applied to determine optimal parameters for the reduction of free fatty acid content of palm kernel oil during its pretreatment for biodiesel production, with a view to reduce the number of experiments and obtaining the mutual interactions among the variables. A (2)4 full factorial central composite design was employed for experimental design and analysis of the results. Four independent variables, viz: methanol to oil ratio, sulfuric acid concentration, reaction time, temperature were tested. All the variables significantly affected the acid value of the product. The optimum values of the four variables were methanol to oil ratio0.255v/v, sulfuric acid concentration 1.591%v/v, reaction time 58.13min, temperature 64.22oC. The corresponding acid value produced was 2.6.After the pretreatment, transesterification reaction was carried out with 6:1 methanol to oil ratio and 0.75w/v KOH to produce biodiesel.The fuel properties biodiesel so obtained were compared with the standard biodiesel.The goodness of the fitness of the model was checked by the determination coefficient R2 which is 0.883 and which indicates that 88.3% of the variability in the response could be explained by the model.

SIMULATION STUDY OF THE PROCESS FOR PRODUCING BUTANOL FROM CORN FERMENTATION

Jiahong Liu, Michael Wang, May Wu
Center for Transportation Research, Argonne National Laboratory, Argonne, IL 60439

Urmila Diwekar
Vishwamitra Research Institute, Westmont, IL 60559

Sept. 17, 2007

Extended Abstract

Acetone butanol ethanol (ABE) fermentation by *Clostridium acetobutylicum* had been the major route to producing butanol prior to 1950s. It started to phase out when the more economical petrochemical routes emerged. Nowadays almost all butanol is produced from petrochemical feedstocks. Research interest has been rekindled in developing viable ABE fermentation processes recently as pursuing non-fossil based feedstock has become a primary alternative to petrochemical feedstocks in reducing the greenhouse gas emissions. In the past 20 years, experimental and computational scientific and engineering attempts to improving ABE fermentation process efficiency entailed various aspects of the process. These attempts include developing new fermentation strains to improve product yield; designing new schemes to minimize butanol inhibition through new fermentor configurations, new downstream processing (production isolation and purification) methods and flowsheeting, and integration of fermentation and downstream processing. For instance, Huang and Ramey et al. (2004) reported an experimental process that uses continuous immobilized cultures of *Clostridium tyrobutyricum* and *Clostridium acetobutylicum* to maximize the production of hydrogen and butyric acid and convert butyric acid to butanol separately in two steps. This process produced butanol at productivity of 4.64 g/L/hr and utilizes 42% glucose, in comparison to the up to 25% glucose utilization in traditional ABE fermentation by *Clostridium acetobutylicum* alone. Qureshi and Blaschek (1999) and Parek, Formanek, and Blaschek (1999) published their experimental and pilot-scale ABE fermentation processes by *Clostridium beijerinckii*. With various downstream processing schemes they developed, glucose utilization can reach 95.1% (Ezeji, Qureshi and Blaschek, 2004). Some of recent studies on integration of fermentation and downstream processing have also shown promising results (Qureshi and Blaschek, 2000; Qureshi and Blaschek, 2001; Qureshi and Blaschek, 2001; Ezeji, Qureshi and Blaschek, 2004; Ezeji, Karcher, Qureshi and Blaschek, 2005; and Qureshi, 2006). An exhaustive survey specifically on major research results for ABE downstream processing can be found in Liu (2003).

Computer simulation that is commonly used in process engineering to scale up experimental results to provide meaningful predictions on behaviors of commercial plant has also been adopted in developing bio-butanol production processes. The earliest effort on ABE fermentation downstream processing simulation were reported in Marlatt and Datta (1986), and Dadgar and Foutch (1988), based on which the economics of their processes were evaluated. More recent studies were published in Liu (2003) and Liu et al. (2004 and 2006). In these studies, downstream processing systems are synthesized, simulated, and optimized in terms of cost. Comparatively, there are few publications on ABE fermentation process simulation.

The current study is concerned with the simulation of ABE fermentation and downstream processing using Aspen Plus® (Aspen Technologies Inc, Cambridge, MA). The simulation is based on the research results from a pilot-scale ABE fermentation plant (Ezeji et al., 2004) and from literatures (Liu, 2003; Liu et al., 2004, 2006; Heckl et al., 2006). Production capacity of the simulated plant is set at 150,000 metric ton/year butanol. The operation runs 315 days a year. Product and byproduct purity specifications are: butanol 99-99.5 wt%; acetone purity 99 wt%; ethanol purity higher than 99.5 wt%. Butanol, acetone, and ethanol are produced by a hyper-butanol-producing strain *C. beijerinckii* BA101 that is able to tolerate toxicity resulted from high butanol concentration. Extensive studies and pilot testing have been performed to characterize this strain (Parekh, M. J. Formanek and H.P. Blaschek, 1999). Corn is fed into a conventional corn dry mill for conversion to glucose through liquefication and saccharification. The glucose is fermented to ABE through a fed-batch system with recycling of fermentation gaseous CO_2 and H_2. After fermentation, acetone, butanol and ethanol are removed by in-situ gas stripping using fermentation gas mixture of CO_2 and H_2. Products are recovered through molecular three-stage distillation and sieve adsorption that separates acetone, butanol, ethanol, and water. Solids and biomass that are removed from fermentor undergo centrifugation and drying. Excessive CO_2 and small amount of H_2 gases are vented to the atmosphere. DDGS (distillers dried grains and solubles) generated from fermentation and distillations are used as animal feed.

Fig. 1 shows the major steps in this segment. As mentioned earlier, we use data derived from this part of corn ethanol process as if it is for the ABE fermentation process except for capacity scaling from 40 million gallon ethanol (MMGY) per year to 150,000 metric tons of butanol per year in the current work. Corn brought into the plant site is first separated from finer particles and foreign objects using a blower and screens. The cleaned dry corn is ground in a hammer-mill and weighed to control the feed rate to the process. The ground corn is then mixed with water, thermo-stable alpha-amylase, ammonia, and lime in a slurry tank. A steam injection heater is used to gelatinize starch and the system is held at pH 6.5 and for 60 minutes at 88°C with agitation. This step is called liquefaction. The starch is then hydrolyzed by alpha-amylase into oligosaccharides. Slurry viscosity sharply increases in the gelatinization step and then rapidly decreases during the hydrolysis. The stream is then "cooked" and held at 110°C for 15 minutes, and transferred to the saccharification tank where the oligosaccharides are converted by glucoamylase to glucose. First, the pH in this tank is lowered to 4.5 by sulfuric acid and the slurry stands for 5 hours. Then glucoamylase is added at 0.11% (dry base), and the oligosaccharides are hydrolyzed into glucose at a temperature of 61°C. During this step, almost all of the oligosaccharides are converted to glucose. The slurry is then cooled to 35°C and transferred to the fermenter and introduced with glucose concentration at 500 gram/liter to the oxygen-free fermentation vessel where glucose has already been inoculated with *C. beijerinckii* BA101. The temperature is controlled at 35 °C and no agitation or pH adjustment is applied during the process. The fermentation proceeds for

22 hours when ABE reaches about 5 gram/liter and then gas stripping is applied. In the condenser above the fermentor, the ABE vapors are cooled to 10 °C and condensed to liquids, leaving uncondensed carbon dioxide and hydrogen in the vapor phase, which are sent through the carbon dioxide scrubber prior to venting. The gas stripping captures most of the butanol, acetone, and ethanol produced during fermentation. The selectivity of the stripping for butanol, defined as

$$[y/(1-y)]/[x/(1-x)]$$

where x = wt% of butanol in fermentation broth
y = wt% of butanol in condensate

is set at 20 (Qureshi and Blaschek, 2001). Accordingly, about 25 wt% water is present in the condensate. It will be removed from the product and byproducts in the downstream processing stage.

The extent of fermentation conversion is set according to experimental and process data. Glucose utilization is 95.1%, i.e., 4.9% glucose is converted into other solids (yeast cells). Butanol, acetone, ethanol, acetic acid, and butyric acid yields are 0.303, 0.155, 0.0068, 0.0086, 0.0084 gram/gram glucose, respectively. The liquid fermentation broth from the condenser is subject to a series of distillation operations where, based on their volatilities, the stream is separated into product 99 wt% pure butanol, 99 wt% pure acetone, and 99.5 wt% pure ethanol. In a mixture containing butanol, ethanol, and water, homogeneous ethanol-water azeotrope and heterogeneous water-butanol azeotrope are formed. In the descending order of their volatilities, the components in the stream, including the two azeotropes, to be separated by the downstream processing are: acetone, azeotrope of ethanol and water, ethanol, azeotrope of water and butanol, water, and butanol. The components in this sequence can possibly or plausibly be separated simply by fractional distillation, fulfilled by various simple and/or complex column configurations, into any two or more subsequences of components at the adjacent components whose volatilities differ sufficiently. The fermentation broth is first fed to the distillation tower where a composition cut is made between the three components, i.e., acetone, ethanol and water, which are withdrawn in the top stream, and component butanol together with trace amount of water, which goes to bottom stream as product with butanol purity 99 wt% that is specified for transportation fuel use . The top stream is sent to the second distillation tower, where acetone is concentrated, purified and then sent out from the top. The bottom stream contains mainly ethanol and water. It is subject to further separation in the third distillation tower where the azeotropic vapor of 94.4 wt% ethanol and 5.3 wt% water (and 2 wt% acetone) is withdrawn from the top and sent to an adsorption unit. Water stream from the bottom is sent back to the fermentor. Water in the top stream is adsorbed in this unit, ethanol purity in the product stream reaches 99.5%, above which the purity of ethanol is required for ethanol to be blended with gasoline and used in vehicles. The adsorption unit consists of two adsorption columns that run cyclically between the adsorption and desorption phases. Molecular sieves are packed in the column as the adsorbents. They are composed of a microporous substance, designed to separate small polar molecules from larger non-polar ones via a sieving action. Water molecules are trapped and adsorbed inside the micro-porous beads, while the ethanol molecules flow around them. Molecular sieves are then regenerated by heat and carrier air. Energy use for the regeneration of the molecular sieve adsorbents is incorporated into the simulation by a Fortran subroutine. Table 1 summarizes the parameters and results. The heat use for the regeneration of adsorbents is calculated as the total amount required to bring the system to a temperature of 500 °F, which includes the adsorption column, the molecular sieves, adsorbed water, and the carrier gas. Cooling of the system is fulfilled by exchanging the heat with the process streams fed to the distillation towers. Therefore, the energy used to cool down the system is not taken into account. The weight of water that the molecular sieves adsorb is set at 22% of the weight of the molecular sieve. The peak concentration of water in the carrier air is set at 1.7 wt%. The adsorption and regeneration cycle time is set at 8 hours. The amount of heat that is required to bring the column to the specified temperature is calculated by the method based on logarithmic mean temperature difference (LMTD) and the scaling of experimental data. The radiation loss factor is set at 5%.

Fig. 2 shows the Aspen output of the process flowsheet from fermentation to downstream processing. Data for the simulation of fermentation and gas stripping unit operations are based on the research results from Blaschek et al. at University of Illinois, Urbana-Champaign on their ABE fermentation facility. The main parameters for fermentation and gas stripping input to Aspen simulation are summarized in Table 1.

The downstream processing flowsheet in the simulation is based on the most cost-effective flowsheet rigorously generated from an exhaustive list of plausible processing equipment and unit operations by Liu et al.'s work (Liu, 2003; Liu et al., 2004, 2006; Heckl et al., 2006). According to their study, this optimal flowsheet consists of a gas stripper that isolates a liquid product stream from the fermentation broth; an adsorption unit that removes majority of water; and a set of distillation columns that purify products butanol, ethanol, and acetone. Their study shows that the cost of this flowsheet is at least 12.5% lower than that of any other alternative flowsheet.

The current simulation integrates the new fermentation and gas stripping unit operations into the existing downstream processing system. Shortcut methods that highlight the mass and energy balances are chosen for simulating most of the unit operations in the process.

Based on the baseline process simulation, various key parameters such as the-state-of-the-art of unit operations, their key process parameters, downstream processing schemes, and their effects on the overall energy consumption of the process, are discussed and compared. The simulation can be used as one of the bases for assessing the potential benefit in energy use and greenhouse gas (GHG) emissions, or the lack thereof, in exploiting corn to produce butanol and utilizing butanol as a potential transportation fuel. The result of this process simulation can provide information in evaluating the viability and potential energy and environmental benefits of butanol as a transportation fuel.

References

- Qureshi, N. and Blaschek, H. P., ABE production from corn: a recent economic evaluation, Journal of Industrial Microbilogy & Biotechnology, 2001, 27, 292-297.
- Qureshi, N. and Blaschek, H. P., Economics of butanol fermentation using hyper-butanol producing clostridium beijerinckii BA 101, Trans IChemE, 2000, Vol. 78, Part C, 139-144.
- Qureshi, N. and Blaschek, H. P., Recovery of butanol from fermentation broth by gas stripping, Renewable Energy, 2001, 22, 557-564.
- Ezeji, T. C., Karcher, P. M., Qureshi, N. and Blaschek, H. P., Improving performance of a gas stirpping-based recovery system to remove butanol from Clostridium beijerinckii fermentation, Bioprocess Biosyst Eng, 2005, 27, 207-214.
- Parekh, M., Formanek, J., and Blaschek, H. P., Pilot-scale production of butanol by Clostridium beijerinckii BA 101 using a low-cost fermentation medium based on corn steep water, App Microbiol Biotechnol, 1999, 51, 152-157.
- Marlatt, J. A., and R. Datta, Acetone-butanol fermentation process development and economic evaluation, Biotechnology Progress, 1986, 2, 23-28.
- Dadgar, A. M., adnd G. L. Foutch, Improving the acetone-butanol fermentation process with liquid-liquid extraction, Biotechnology Progress, 1988, 4, 36-39.
- Huang, W.C., Ramey, D. E. Yang, S. T., Continuous production of butanol by Clostridium acetobutylicum immobilized in a fibrous bed bioreactor, Applied Biochemistry and Biotechnology - Part A Enzyme Engineering and Biotechnology, 2004, 113-116, 887-898.
- Ezeji, T. C., et al. Acetone butanol ethanol (ABE) production from concentrated substrate: reduction in substrate inhibition by fed-batch technique and product inhibition by gas stripping, Appl Microbiol Biotechnol, vol. 63, 2004, 653-658.
- Liu, J., Downstream Process Synthesis for Biochemical Production of Butanol, Ethanol, and Acetone from Grains, PhD Dissertation, 2003.
- Liu, J., L. T. Fan, P. Seib, F. Friedler, and B. Bertok, Downstream Process Synthesis for Biochemical Production of Butanol, Ethanol, and Acetone from Grains: Generation of Optimal and Near Optimal Flowsheets with Conventional Operating Units, Biotechnol. Prog., Vol. 20, pp. 1518-1527, 2004.
- Liu, J., L. T. Fan, P. Seib, F. Friedler, and B. Bertok, Holistic Approach to Process Retrofitting: Illustration with Downstream Process for Biochemical Production of Organics, Ind. Eng. Chem. Res., Vol. 45, pp. 4200-4207, 2006.
- Quresh, N., 2006, Personal communications with May Wu of Argonne National Laboratory.

Tables and Figures

Table 1. Main input parameters of fermentation and gas stripping for Aspen simulations

Input parameter		Value used in the simulation	Value range reported in the references*
Fermenter operating temperature (°C)		35	30 ~ 67
Fermenter operating pressure (atm)		1	n/a
Glucose utilization in fermentation (g/l)		500.1	45.4 ~500.1
Yield (gram/gram Glucose)	Acetone	0.155	n/a
	Butanol	0.303	n/a
	Ethanol	0.0068	n/a
	ABE (total)	0.465	0.27 ~ 0.47
	Acetic acid	0.0086	n/a
	Butyric acid	0.0084	n/a
	Hydrogen	0.021	n/a
	Carbon dioxide	0.6954	n/a
ABE productivity (g/(l.hr))		1.16	0.17 ~1.16
Gas stripping carrier gas		Fermentation gas	Fermentation gas
Gas stripping selectivity for butanol		20.0	4 .0 ~ 30.5
Gas recycle rate (l/(l.min))		3.00	0.67 ~ 10.00
Condensation temperature (°C)		2	-60 ~ 4

*References: Ezeji et al. (2004); Qureshi et al. (2001).

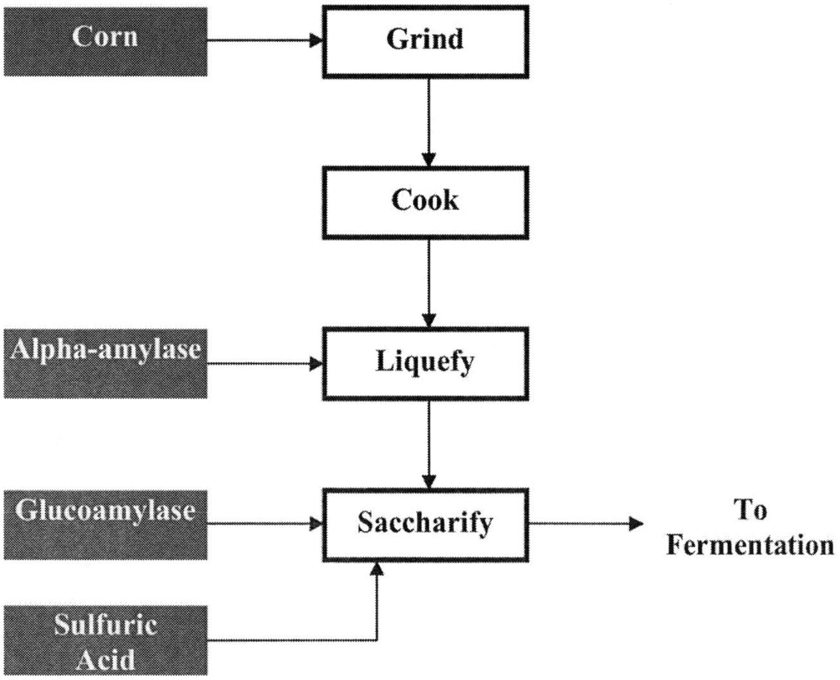

Fig. 1. Schematic for process of grain receiving, liquefaction, and saccharification.

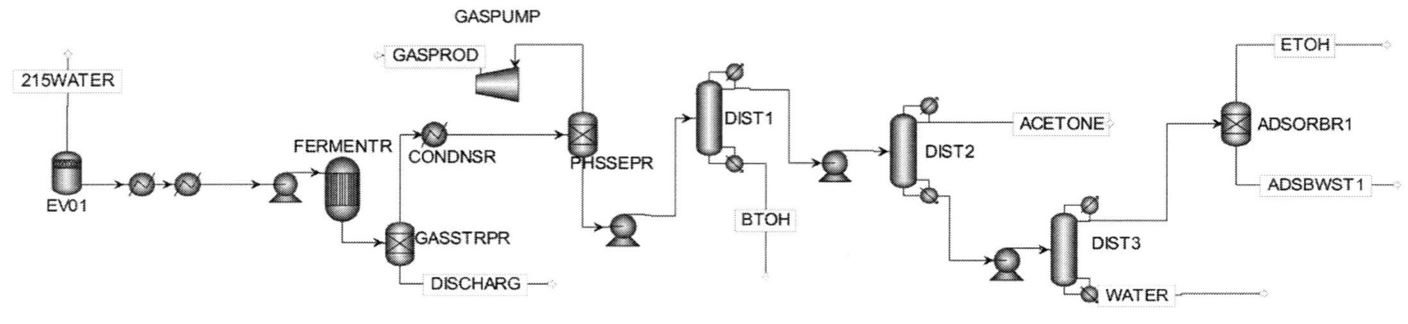

Fig. 2. Aspen Plus Baseline Process Flowsheet from Fermentation to Downstream Processing.

Coal Characterization in CBM/ECBM Processes Using X-ray CT Analysis

R.Guo[1], Y. Bao[2], Y. Li[3], A. Kantzas[1,3]

1. Department of Chemical & Petroleum Engineering, University of Calgary, AB, Canada
2. Department of Computer Science, University of Calgary, AB, Canada
3. Tomographic Imaging and Porous Media Laboratory, Calgary, AB, Canada

Introduction

Coalbed methane (CBM) is an unconventional natural gas resource that gains increasing worldwide attention. Canada has vast resources of coal and it has been estimated that the total in-place reserves are 36 x 10^{12} m^3. Over 60% of Canada's CBM resource is in Alberta (1). Coalbed methane has the potential of contributing a significant portion of Canadian natural gas production in the foreseen future. The primary CBM production mechanism is to recover the methane gas by reservoir pressure depletion. The primary CBM production usually has less than 50% of the recovery rate. To further improve the gas recovery rate, CO_2 is proposed to be injected into the CBM reservoir in recent years. This enhanced methane recovery mechanism has been discussed by several researchers (2, 3). If successful, its implications include CO_2 sequestration in deep unmineable coalbed formations.

Coal seams are heterogeneous in terms of lithotypes and morphologies. This creates a challenge in understanding the subsurface behavior of the coal and injected gas during primary and enhanced gas recovery processes. As an organic rock, coal structure is easily deformed by the net stress imposed on it. Therefore, coal experiences many changes in stress conditions during the production life of the CBM reservoir. Because the network of natural fractures and cleats in a coal determines to a large extent the mechanical properties of the coal. Coal is very soft and has low elastic modulus compared to the rock formation. The stress and time dependent deformation of the coal porous structure is expected to change the behavior of the most important properties of the coal, such as porosity and permeability, which in turn change the reservoir production predictions. Coal physical properties such as density and compressibility are also dynamically changed with the coal structure deformation.

Computerized tomography (CT) is an imaging technique that generates an image of a thin, cross-sectional slice of a test piece. It provides a non-destructive method that visualizes the internal structure of an object by reconstruction of a matrix of linear attenuation coefficients. Since x-ray CT allows continuous non-destructive observation of the sample during the experiment, it was proven to be a valuable tool in petroleum exploration and development research (4, 5, 6, 7). Although X-ray CT techniques are used extensively for reservoir rock characterization as well as fluid saturation monitoring during flow experiments, application to coal deformation studies has been limited to a small number of studies (8, 9, 10, 11).

Verhelst et al. (8) used x-ray to image coal density. They correlated three-dimensional tomodensity (or linear attenuation coefficient for x-rays) measurements with two-dimensional data on the composition of a coal sample acquired with color image analysis (CIA). Pyrak-Nolte et al. (9) used x-ray CT to obtain the fracture aperture size and spatial distributions of bituminous coal cores from Fruitland Formations.

Karacan and Okandan (10) observed the gas storage and transport in a coal sample by x-ray CT. They conducted gas flow experiments on a 3.8 cm diameter core plug of coal from Turkey. They investigated the effect of different coal structures on the gas transport and storage and the matrix/fracture interactions using x-ray CT coupled with scanning electron microscopy (SEM). They also determined the diffusion coefficients from the gas recovery data calculated from CT images.

The behavior and effect of different coal lithotypes during sequestration of CO_2 into coal seams are important knowledge gaps when modeling sequestration processes and planning for field application. Mathews et al. [12] studied the CO_2 sorption with time within the coal samples using x-ray CT techniques. Karacan and Mitchell [11] presented the laboratory study for sequestration of CO_2 in a 2.5 cm diameter Pittsburgh coal sample kept under a constant effective stress during gas uptake. CO_2 storage was observed by x-ray CT scanning. They showed that different microlithotypes behave differently during CO_2 adsorption. Vitrite was observed to swell and to decrease the initial density about 5%, when CO_2 diffused into the micropores. A density increase was observed in other microlithotypes present in the sample due to gas adsorption.

X-ray CT techniques are proposed as tools to characterize coal in the CBM primary and enhanced gas recovery processes. The purposes of the x-ray CT experiment in this work are to characterize coal and demonstrate the use x-ray CT to investigate the coal density, and density distribution varied with the gas type and gas pressure.

Experimental Work

Sample Preparation

The coal sample comes from the Manville Group in Alberta. Table 1 lists the characteristics of the coal sample. The depth of the coal seam is 853 m. Prior to starting the experiments, the coal core sample was kept in a moisture oven at 25°C, to simulate reservoir conditions. Proximate analysis was carried out on sub-samples of the crushed coal to determine moisture (ASTM D3173), ash (ASTM D3174), and volatile matter (ASTM D3175) contents [13]. Table 1 also shows the results of the proximate analysis and vitrinite reflectance measurements. The moisture and ash content of the coal are 9% and 14.1% respectively. Vitrinite reflectance is a well accepted means of determining coal rank. According to the coal rank, the rank of this piece of coal is sub-bituminous B indicated by its 0.47% vitrinite reflectance.

Unlike reservoir rock core samples, coal core usually has irregular cross-section area. This makes it difficult to mount into a core holder directly. The mechanical properties of coal are also dramatically different from those of sandstone or carbonate rocks. Coal is friable. The overburden stress load may cause permanent coal structure change. Therefore, the coal core was encased in urethane resin in a cylindrical mold. Then the resin coated coal core was sealed in a rubber sleeve inside an x-ray transparent high-pressure core holder. The urethane is solid enough to keep the core intact and confine it in the rubber sleeve, but retains enough flexibility to transmit overburden pressure to the coal. The coal core had an average cross-sectional area of about 3580 mm^2 and was 85.5 mm long.

Table 1: Proximate analysis and coal rank classification

Table 1. Proximate analysis and coal rank classification						
Coal type	Depth (m)	Moisture(*) (%)	Ash(*) (%)	Vol. Matter (%)	Vitrinite reflectance (%)	Rank
Deal Bruce	853	9	14.1	36	0.47	Sub-B
* Equilibrium moisture basis.						

Core-flood Experiments

The core flood experiments are consisting of helium, methane, CO_2 flow, gas adsorption and desorption, and the displacement of methane by CO_2. First, helium as a non-adsorbing gas was flowed through core at different combinations of overburden and pore pressure at a temperature of 23°C. The pore pressure (core outlet pressure) was fist kept at atmospheric pressure and overburden pressures increased. Therefore the net stress on the core (defined as the difference between overburden and average pore pressure) increased. The pore pressure was then increased at constant overburden pressure, thus decreasing the net stress on the core. Then the core was saturated with methane at a given system pressure. After adsorption was reached, methane flowed through the core at the given system pressure. Then the system pressure was reduced to produce methane. The amount of methane desorbed was measured. This process is called the primary CBM production. After CBM production by pressure depletion, CO_2 was injected to displace methane. This is so called CO_2-ECBM process. At the equilibrium CO_2 saturation conditions, CO_2 flowed through core at several system pressures. At the end of ECBM, the core was blown down and the CO_2 was allowed to desorb. Helium was flowed through core again to determine the permanent changes on the core. Figure 1 represents the schematic of the core-flood rig.

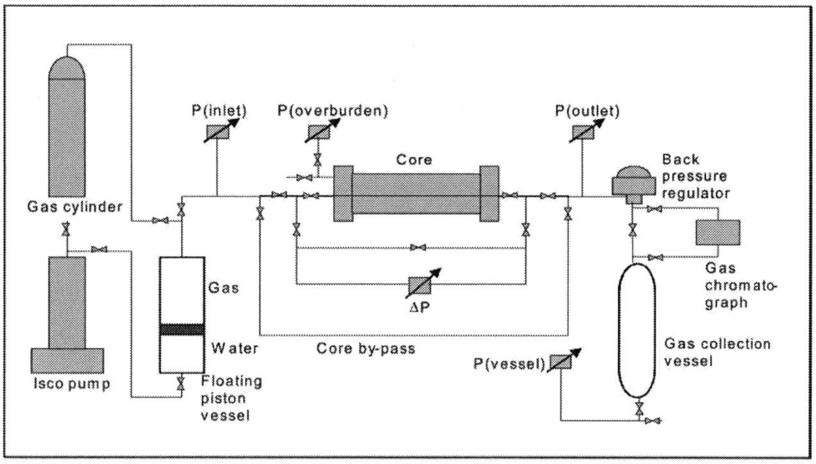

Figure 1: Schematic of the core-flood rig

CT Experiments

The x-ray CT experiments were carried out parallel to the core flood experiments to provide x-ray images of coal core saturated with different gases at different system pressures. The CT scanner used in all the CT experiment is the 3rd generation GE9800 (Figure 2). In total, there were 17 CT experiments conducted during helium flow, methane saturation and flow, CO_2 adsorption and desorption stages. Table 2 lists the pressure conditions of each CT experiment in order. All the experiments were carried out in the room temperature.

Prior to each CT experiment, the scanner was properly warmed up and calibrated. Calibration scans of air, or uniform objects called phantoms, provide the baseline information for the system to produce object images. The system needs calibration data for every possible combination of scan parameters. The parameters used for the CT scan are summarized in the Table 3. The selection of the scanning parameters is based on the property of the scanned material for best image contrast and resolution. Consistent scanning parameters were used for all the CT experiment to produce comparable CT images.

Table 2: CT experiment conditions

Gas type	Gas pressure					
	Overburden P_{ob}		Pore P_{pore}		Net Overburden P_{nob}	
	psig	MPa	psig	MPa	psi	MPa
Air	0	0.09	0	0.09	0	0
Helium	490	3.47	0	0.09	490	3.38
(before CBM)	1517	10.55	0	0.09	1517	10.46
	1517	10.55	520	3.67	997	6.88
	1551	10.78	1020	7.12	531	3.66
Argon	996	6.95	52	0.45	944	6.51
Methane	1529	10.63	805	5.64	724	4.99
	988	6.90	320	2.29	668	4.61
CO_2	1004	7.01	309	2.22	695	4.79
	1541	10.71	313	2.25	1228	8.46
	1527	10.62	505	3.57	1022	7.05
	1468	10.21	811	5.68	657	4.53
	1017	7.10	299	2.15	718	4.95
Helium	1000	6.98	110	0.85	890	6.13
(after ECBM)	533	3.76	0	0.09	533	3.67
	120	0.92	0	0.09	120	0.83
	0	0.09	0	0.09	0	0.00

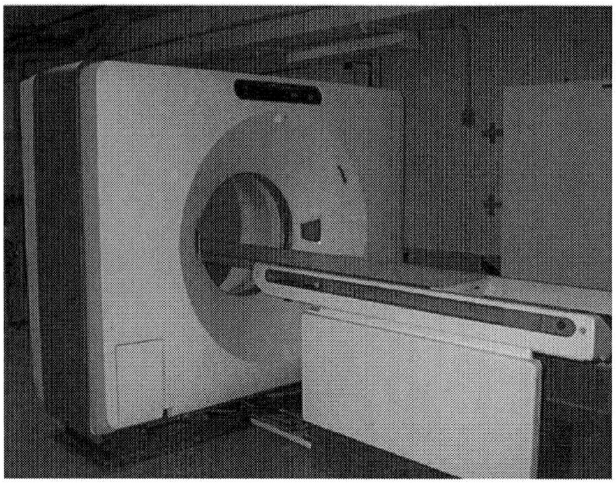

Figure 2: X-ray CT scanner GE9800

X-Ray CT Image Processing

The raw data obtained from each CT scans is a CT image containing 512 × 512 pixels. For DFOV of 10 cm, each pixel stands for 0.2 mm ×0.2 mm ×1.5 mm volume element. Each pixel readings as CT number, ranging from -1024 to +3071, is an average CT number within that volume and is a function of mean density and effective atomic number.

The first step to process the raw data images is to remove the surrounding resin from coal. The resin has CT numbers which are less than those of coal matrix. A threshold CT number was chosen to remove resin from the coal boundary. The second step of image processing is the mapping of coal bulk density using density calibration:

$$\rho = aCT_n + b \tag{1}$$

Where ρ is the density; a is the slope; b is the intercept; and CT_n is the CT number. The slope a and intercept b can be obtained through calibration using reference materials with known bulk densities.

The result of the density calibration is in Figure 3. The CT numbers were collected to generate the statistics study on coal property information. We obtained average, maximum, and minimum CT numbers (include coal matrix and fracture) as well as the average coal matrix and fracture CT numbers for each slice.

Table 3: CT scanning parameters

Scan parameter	Scan type	
	Axial	Scout
Voltage (kV)	120	120
Filament current (mA)	120	70
Scan spacing (mm)	1.5	–
Scan thickness (mm)	1.5	–
Scan time (s)	3	–
SFOV (cm)	25	–
DFOV (cm)	22, 10	–
Matrix	512 x 512	–
Algorithm	Detail	–

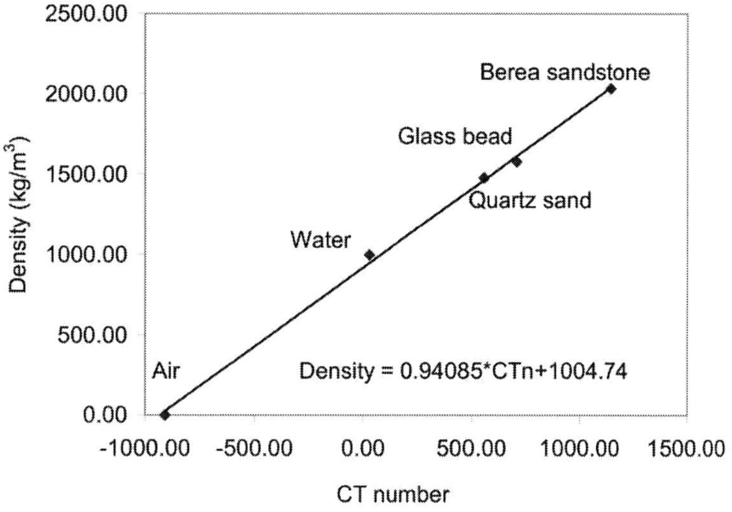

Figure 3: Density calibration

Results and Discussion

Density Logging

Once we transform the CT numbers into the coal bulk density for each scan location, we can plot the coal average bulk density change with the core location at different flow conditions. Figure 4 shows the coal bulk density log for helium flow. The lower group of curves in the graph, marked as group 1 is for helium flow before CBM primary production. The upper group, group 2, is for helium flow after ECBM processes. Since helium is a non-adsorbing gas, the density variation demonstrated here is contributed from the pressure effect. The lowest solid blue symbol and line in Figure 4 is for the original core without any pressure. The result shows that coal core is a highly heterogeneous material. Its density varies dramatically at different locations. The lowest density is as low as 1130 kg/m^3, which corresponds to the largest fracture region. While the highest density reflects the mineral streaks, its density is over 1200 kg/m^3. The range of the density is typical for coal. Coal bulk density is strongly dependent upon rank [14, 15], pore volume and structure [15], and coal composition. The density of mineral material normally exceeds 2000 kg/m^3. The density of coal macerals ranges from 1200 to 1700 kg/m^3. The density of cleats and pores may be approximately zero in the gas phase. In the group 1 curves, coal bulk density is directly increases with the incremental of the net overburden pressure imposed upon core. The reason is quite obvious because high pressure compresses coal and closes its fracture, hence reducing its volume. The lowest density point in the original core experiences the greatest density decrease due to the high compressibility of the coal fracture. The pressure effect is very consistent for all the pressure settings. The same pressure effect also applies to the group 2 curves. However, the differences between the curves are much smaller than that of group 1 curves. This means that there were the CBM primary and ECBM processes incurred permanent changes on coal structure. Therefore, coal core can not relax into its original status even the pressure load was removed at the final stage. The lowest empty blue symbol and line in the group 2 curves exhibits the final core density profile. Another noticeable point in this graph is that the pressure effect on coal density is much larger at the ends of the core than the middle part.

The density loggings during the CBM primary and ECBM processes are plotted in Figure 5. The lower three curves are for argon and methane, and the rest are for CO_2 at different pressure settings. The net overburden pressures used in the core-flood experiment for methane flow do not cause obvious coal density change due to the pressure difference. In Figure 5, when the net stress changes from 4.99 MPa to 4.61 MPa, two density curves for methane flow overlap with each other. However, the effect of the gas type on the coal density is much pronounced. The first curve for CO_2 (4.19 MPa) corresponds to the core after methane displaced by CO_2. We can see that coal density has a dramatic increase. This effect is due to the higher affinity of CO_2 to coal than methane. It is well accepted fact that CO_2 has more than twice the adsorption capacity than methane. Therefore CO_2 causes much severe swelling effect on coal matrix than methane. The expansion of the coal matrix results in the coal fracture close. Thus the coal density increases. The effect of the gas type on coal density reflects the degree of coal swelling due to gas adsorption. When net overburden pressure increased on core, coal density may increase or decrease. The trend of the pressure effect on CO_2 curves is not clear from this graph.

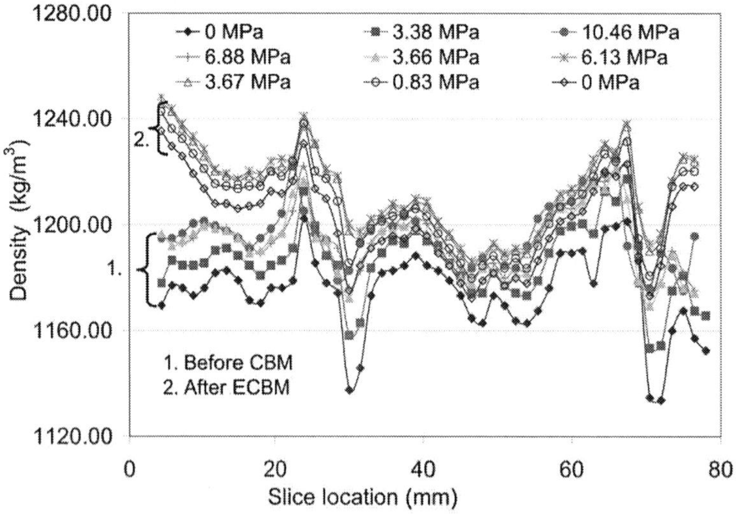

Figure 4: Density log for helium flow at different net overburden pressure

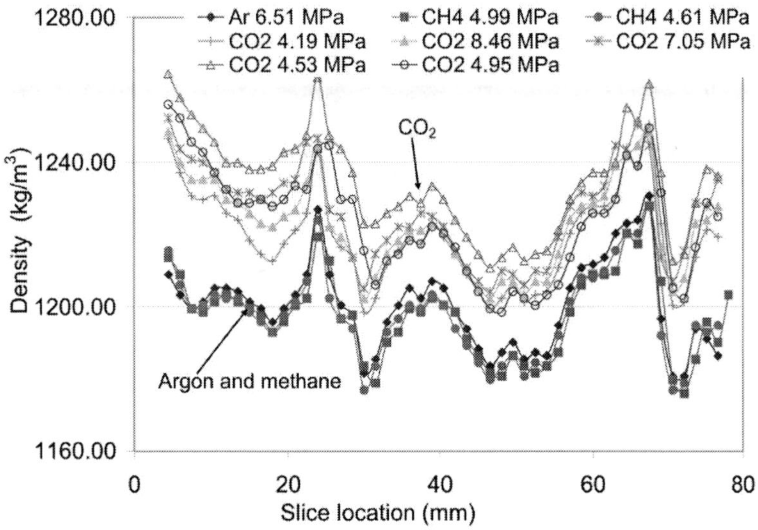

Figure 5: Density log with CO_2 at different net overburden pressure

Density Variation

To further investigate the effect of the pressure and gas type on the coal density, the average bulk density for the whole core was calculated and plotted against the net overburden pressure. Figure 6 represents how coal density changes during the core-flood experiment. First, the coal average density increases with the net stress due the increased overburden pressure. It decreases as the net stress reduces by increasing the pore pressure. Here the observed hysteresis indicates permanent changes in coal structure. Coal has larger density compared to the observed trend when increasing overburden pressure at the same net stress. This hysterisis is consistent with the coal permeability change reported in a previous publication [16]. Introducing methane in coal increases coal average density. The increase is only 1% compared to the increasing trend and 0.4% compared to the decreasing trend of helium flow. After methane production, coal density is boosted by CO_2 adsorption by 1.8% compared to methane at the same pressure level. The coal density change due to the adsorbing gas flow is correlates well to the coal permeability change [16]. Coal permeability is provided by the coal fracture network. The swelling of the coal matrix due to adsorption of gas on coal squeezes the coal fracture, which is the gas flow path. Therefore the coal permeability decreases during methane and CO_2 flow. Hence both coal density and permeability are affected by the same factors and follow the similar trend.

It is very interesting that coal density under CO_2 saturation is not a single function of net stress. It forms a triangle loop with the changes of the net overburden pressure. This implies that another factor may also affect coal density. Since the net overburden pressure in the net difference between the overburden pressure and the pore pressure and only pore pressure affects the gas adsorption capacity, it is necessary to isolate the effect of the pore pressure on the coal density. In Figure 7, the coal density change as a function of the pore pressure at the constant overburden pressure is demonstrated. For helium flow, pore pressure increase results in a slight density decrease. The correlation is linear in nature. This is because the increasing pore pressure can partly open coal fracture spaces, leading to coal volume increases. On the contrary, coal density increases with the pore pressure during CO_2 flow at the constant overburden pressure. Regression yields a slope as large as 3.5. This explains why coal density is not a single function of net stress in CO_2 flow. Because coal density increases as the net stress decreasing after 8.46 MPa in Figure 5, the effect of CO_2 adsorption on coal density overcomes the effect of the gas pressure.

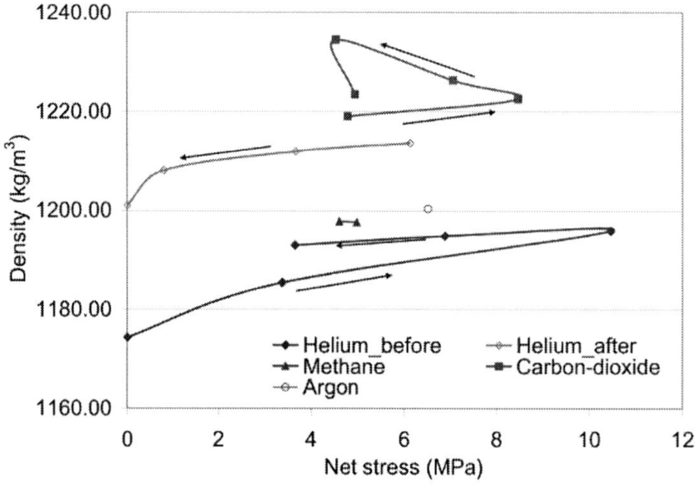

Figure 6: Coal average bulk density change with gas type and net overburden pressure

Conclusions

From the above results, we can see that net overburden pressure, gas adsorption capacity, and production history are all key factors affecting coal core structure, leading coal density and density distribution variations. Hence, the core flow path, which contributes to the coal permeability, changes with those factors during CBM and ECBM processes. The swelling of coal matrix due to adsorbing gases has greater effect on coal density than imposing pressure. This study shows that x-ray CT technique is a useful tool in coal characterization during CBM/ECBM processes.

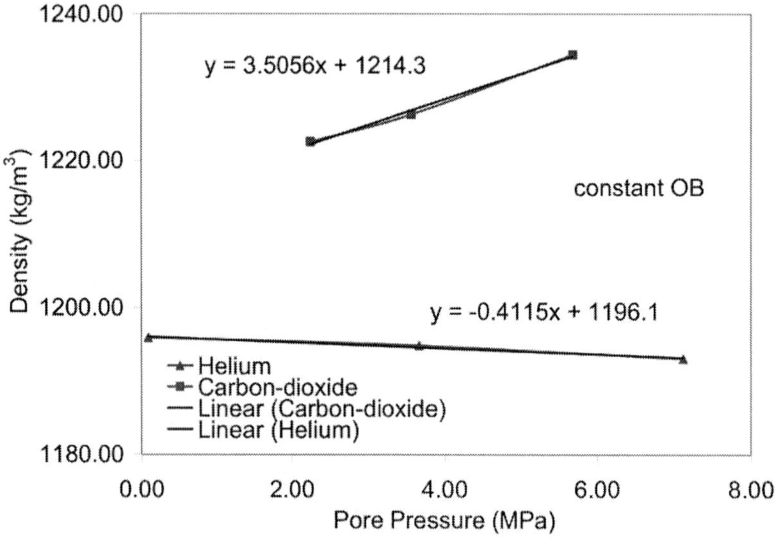

Figure 7: Coal bulk density change with coal pore pressure

Acknowledgement

The authors would like to thank the Natural Sciences and Engineering Research Council of Canada, The Canada Research Chair in Energy and Imaging, its industrial affiliates (Nexen, Canadian Natural, Shell, Petro-Canada, E-T Energy, Devon Canada, Schlumberger, Suncor, Paramount, CMG, ConocoPhillips and Laricina Energy), Dominion Exploration Canada, and Imperial Oil Resources for funding and support. The contributions from the staff of the Tomographic Imaging and Porous Media Laboratory are greatly appreciated.

References

1. Anderson, JR., et al., (2003), "Producing natural gas from coal", *Oilfield Review*; 8-31, Autumn.
2. Seidle, J.P., (2000), "Reservoir engineering aspects of CO_2 sequestration in coals", *SPE 59788, SPE/CERI Gas Technology Symposium*, Calgary, Alberta, 3-5 April.
3. Wong, S., Gunter, W.D., Mavor, M.J., (2000), "Economics of CO_2 sequestration in coalbed methane reservoirs", *SPE 59785, SPE/CERI Gas Technology Symposium*, Calgary, Alberta, 3-5 April.
4. Wellington, S.L. and H.J. Vinegar, (1987), "X-Ray Computerized Tomography", *Journal of Petroleum Technology*, Aug, 885-898.
5. Kantzas, K., (1990), "Investigation of physical properties of porous rocks and fluid flow phenomena in porous media using computer assisted tomography", *In Situ*, 14(1), 77-132.
6. Carlson, W.D., Rowe, T., Ketcham, R.A., and Colbert, M.W., (2003), "Applications of High-Resolution X-Ray Computed Tomography in Petrology", *Metroritics and Palaeontology*, Geological Society, 215, 7-22.
7. Akin, S., and Kovscek, A. R., (2003), "Computed Tomography in Petroleum Engineering Research, Geological Society", 215, 23-38.
8. Verhelst, F., David, P., Fermont, W., Jegers, L., Vervoort, A., (1996), "Correlation of 3D-computerized tomographic scans and 2D-colour image analysis of Westphalian coal by means of multivariate statistics", *International Journal of Coal Geology*, 29, 1-21.
9. Pyrak-Nolte, L. J., Montemagno, C.D., and Nolte, D.D., (1997), "Volumetric imaging of aperture distributions in connected fracture networks", *Geophysical Research Letter*, 24(8), 2343-2346.
10. Karacan, C. O., and Okandan, E., (2001), "Adsorption and gas transport in coal microstructure: investigation and evaluation by quantitative X-ray CT imaging", *Fuel*, 80(4), 509-520.
11. Karacan, C. O., Mitchell, G. D., (2003), "Behavior and effect of different coal microlithotypes during gas transport for carbon dioxide sequestration into coal seams", *International Journal of Coal Geology*, 53(4), 201-217.
12. Mathews, J.P., Karacan, O., Halleck, P., Mitchell, G. D., (2001), "Storage of pressurized carbon dioxide in coal observed using x-ray tomography", *First National Conference on Carbon Sequestration*, Washington, DC, United States, May 14-17.
13. American Society for Testing and Materials, Annual Book of ASTM Standards, Vol. 05.05, Gaseous Fuels, Coal and Coke, *American Society for Testing and Materials*, West Conshohocken, Pennsylvania, 1999.
14. Gan, H., Nandi, S. P., Walker, Jr. P. L., (1972), "Nature of the porosity in American coals", *Fuel*, 51(4), 272-277.
15. Levine, J.R., Johnson, P.W., Beamish, B.B., (1993), "High pressure microgravimetry provides a viable alternative to volumetric method in gas sorption studies on coal", *Proc. 1993 Int. Coalbed Methane Symp.* Univ. of Alabama, Tuscaloosa, pp. 187–195.
16. Guo, R., Mannhardt, K. and Kantzas, A., (2007), "Laboratory investigation on the permeability of coal during primary and enhanced Coalbed Methane production", *The 58th Canadian International Petroleum Conference*, Calgary, Alberta, June 12-14.

Nomenclature

a	slope
b	intercept
CT_n	the CT number
ρ	bulk density
CBM	coalbed methane
CIA	color image analysis
CT	computerized tomography
CO_2-ECBM	CO_2 enhanced coalbed methane

Co-electrolysis of Steam and Carbon Dioxide as Feed to a Methanation Reactor

L. Frost, S. Elangovan, J. Hartvigsen – Ceramatec, Inc

Introduction

Certain natural gas deposits contain a high percentage of carbon dioxide that is co-produced with the methane fraction of the natural gas. Conventional production methodologies involve separation of the carbon dioxide from the methane and release of the carbon dioxide. In the current environment desiring the limiting of greenhouse gas emissions to the atmosphere, this creates potential problems. Ceramatec, a small business located in Salt Lake City, Utah, has a potential solution that can solve this problem. The basis for this solution deals with the use of environmentally benign methods for generation of electricity coupled with electrolysis of carbon dioxide and steam to form synthesis gas (i.e. hydrogen and carbon monoxide). The synthesis gas can be catalytically processed in a number of manners to yield additional methane that can be exported with the natural gas stream. This paper describes the technology and alternative paths to removal of carbon dioxide from the natural gas, processing to convert the carbon dioxide to carbon monoxide, and formation of methane to add to the natural gas.

Technology

The technology centers on the use of the solid oxide fuel cell materials' set operated in an electrolysis mode. By applying an electric current across the cell, it is possible to drive the fuel cell material to transfer oxygen ions in the reverse of the usual direction. The basic equation is:

$$2\ CO_2 + \text{electricity} \rightarrow 2\ CO + O_2$$

By the nature of the fuel cell structure, the oxygen is separated from the carbon monoxide and thus can be available for other processes. The process has been proven at Ceramatec both in independent experiments and in conjunction with the Idaho National Laboratory, a United States Department of Energy laboratory.

The same materials can be used to electrolyze water and produce hydrogen. The reaction for the water electrolysis is:

$$2\ H_2O + \text{electricity} \rightarrow 2\ H_2 + O_2$$

There appear to be certain benefits to co-electrolysis of water and carbon dioxide based on experimental data obtained to date. Shown in the following figures are the construction of a single cell and a laboratory test unit consisting of a number of cells.

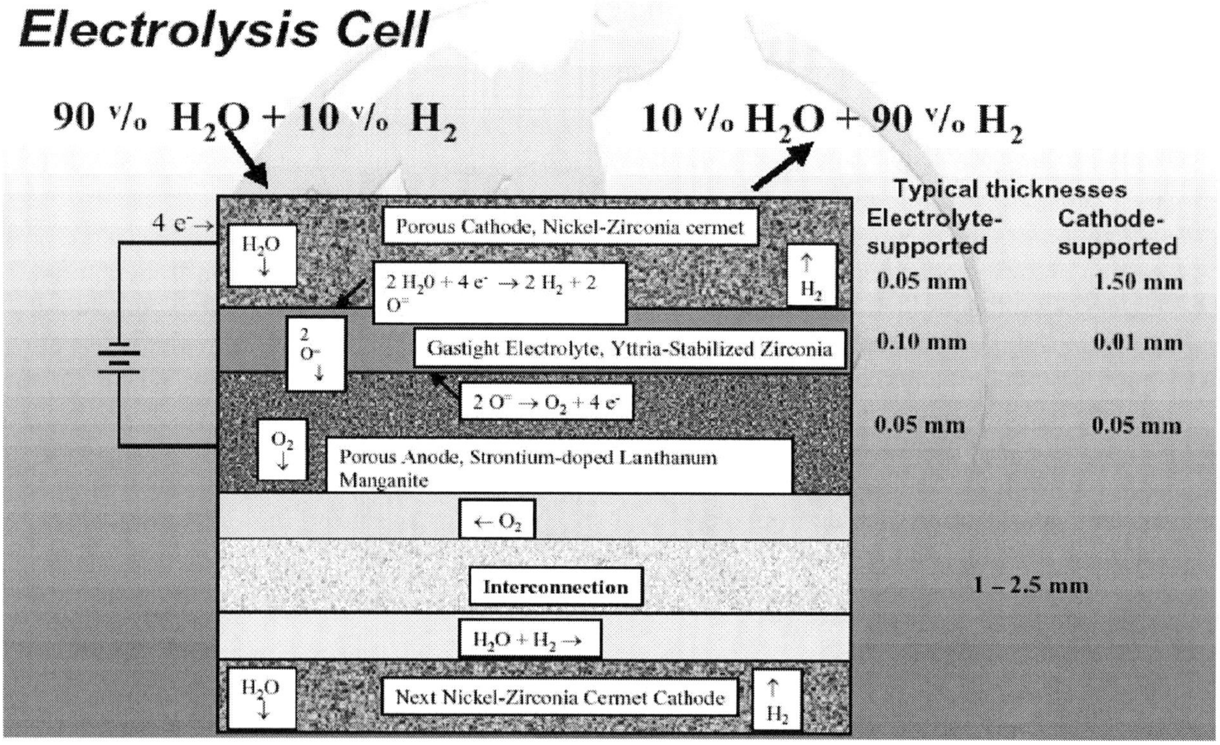

Figure 1: An electrolysis cell (water shown but also applicable for carbon dioxide)

Figure 2: Laboratory scale unit used for testing (10 cells)

The above figures illustrate the technology using the steam electrolysis mode but the same technology can also be used in the electrolysis of carbon dioxide.

Ceramatec and the Idaho National Laboratory are currently constructing a larger scale system that will demonstrate the process with larger stacks integrated into a complete system. An artist's conception of that system is shown in Figure 3 below.

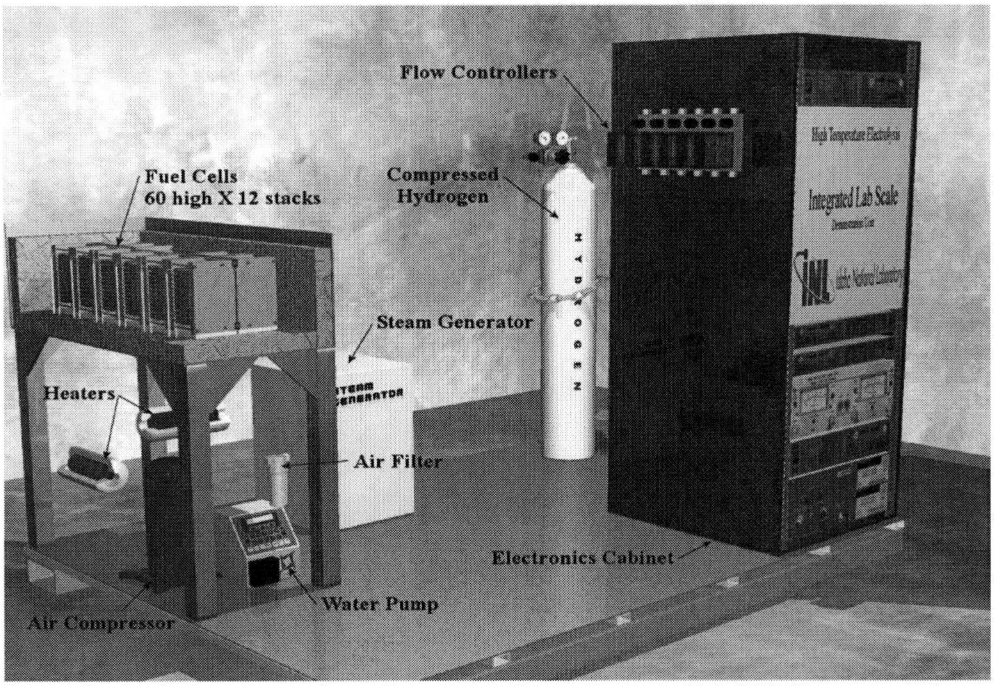

Figure 3: Artist's conception of a steam electrolysis unit (~250 liters per minute)

The steam electrolysis units shown above and the tested carbon dioxide electrolysis units operate at the typical temperatures of a solid oxide fuel cell system (i.e. ~ 800° C). This has the advantage of supplying a portion of the energy required for separating the oxygen from the water or carbon dioxide molecule from heat rather than electricity. The relationship between the heat and electricity input is shown in Figure 4. The overall efficiency is approximately twice that of conventional electrolysis.

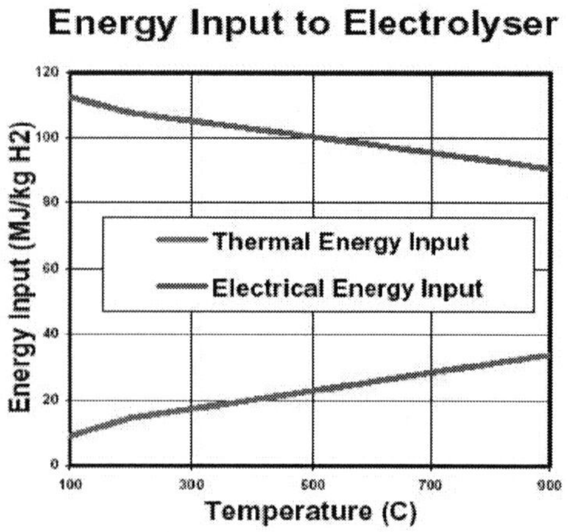

Figure 4: Relationship between electric and thermal energy inputs

As can be seen from the figure as energy is input from thermal, the electrical input required per kilogram of hydrogen drops. In general, the same relationship holds for the disassociation of carbon dioxide.

Proposed use of the technology

There are several locations in the world where natural gas deposits coexist with the availability of renewable resources that can be used to generate electricity. Wind, solar, hydroelectric, biomass, and geothermal are all alternatives that could be used to generate electricity in a carbon neutral manner. The electricity could then be used in conjunction with the above technology to generate synthesis gas from a mixed stream of water (i.e. steam) and carbon dioxide. When the resultant synthesis gas is fed to a methanation reaction, the resultant methane could be recombined with the extracted natural gas to form a relatively pure methane stream. Dependent on the percentage of carbon dioxide, it may be possible to directly feed the methanation reaction with the hydrogen generated and the carbon dioxide without the necessity for a separation step. The water percentages would be adjusted as necessary for the various mixtures of methane and carbon dioxide. A typical representation of the system shown in block diagram form is shown in the figure below.

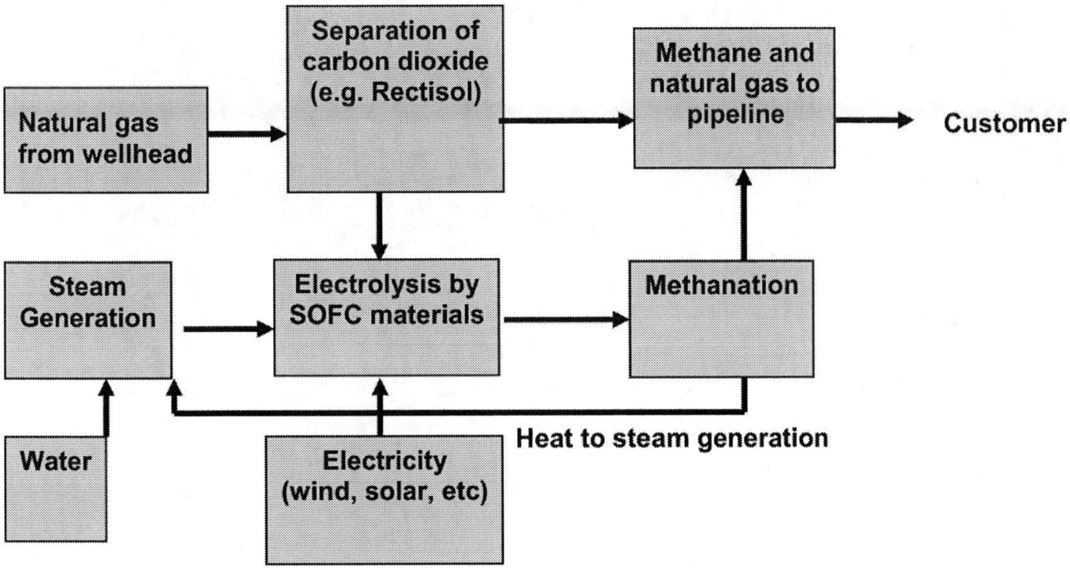

Figure 6: Block diagram of electrolysis of carbon dioxide and water

There are a number of alternative scenarios that might be considered dependent on the conditions at the particular location. It might be possible to avoid the carbon dioxide separation step under certain percentages of carbon dioxide and feed the natural gas directly to the electrolysis system. Care would need to be taken to avoid the formation of coke from reaction with the methane and resultant damage to the electrolysis system but within those constraints this may be an interesting possibility. A few of the possible process variations are listed below:

- Separation of CO_2, co-electrolysis of CO_2, methanation of resultant syngas
- Direct co-electrolysis of CO_2 containing natural gas followed by methanation
- Hydrogen production, added to CO_2 containing natural gas in methanation reactor.

Anticipated Program

It should be recognized that the electrolysis unit has only been demonstrated at a laboratory scale at this point in time. The largest unit tested to date has shown a combined $CO+H_2$ production rate of 1.25 normal m^3/hr, operating for ~1000 hours producing syngas, continuing another ~1000 hours producing only hydrogen. A photo of this unit is shown in Figure 7.

Figure 7. High Temperature Electrolysis Module Rated 1.25 normal m^3/hr H_2+CO

It is critical to be able to demonstrate the ability to scale the electrolysis unit and operate the unit under commercial conditions. This will require some development effort and a corresponding period of time. The potential benefits (i.e. removal of carbon dioxide and generation of additional methane) of a successful program, however, are substantial. As some of the deposits of natural gas are located in foreign countries, it may also engender goodwill in the local area by reducing greenhouse gases, extending natural resources, and potentially involving local research organizations.

The ability of the electrolysis unit to co-electrolyze carbon dioxide and water has been previously demonstrated. Thus, the first suggested step is not to duplicate these results but to investigate the projected types of feed streams that might come from natural gas reservoirs of interest and to model the projected system that would be required for those reservoirs. It is expected that the customer will be highly involved in setting the parameters that define the boundary of the study. Significant input will be required on such items as the carbon dioxide percentages, other impurities, availability of water.

In parallel with that modeling effort, methodologies for increasing the size of the electrolysis system would be investigated. This scale-up would take advantage of other projects that are looking at alternative methods for increasing the size of individual cells and the integration of these cells into an operating system and the overall process. The output expected from this initial phase would be a report on the technical feasibility of proceeding with the development of a pilot system.

Some results have been shown by using the co-electrolysis unit at a laboratory scale and putting the resultant synthesis gas into a small, atmospheric methanation reactor. The results are shown in Table 1 below.

Table 1
Tests for Generation of Methane from CO_2 and H_2O
(numbers are volume percents)

	FROM ELECTROLYSIS			*FROM METHANATION*		
	CH_4	CO	H_2	CH_4	CO	H_2
Test 1	.0	14.3	60.7	42.5	0	13.8
Test 2	.7	18.5	58.0	47.7	0	9.2
Test 3	.3	20.1	63.5	50.0	0	9.4
Test 4	.1	15.8	58.9	42.0	0	4.6
Test 5	.1	15.2	59.5	40.4	0	8.2

As can be seen in the table, all of the carbon monoxide was consumed in the methanation reaction. There was some small amount of methane coming from the solid oxide electrolysis cell. This was most likely formed from the combination of the carbon monoxide and hydrogen in the presence of the metal piping.

394h Combustion Of Off-Stoichiometric AlMoO3 Nano Composite Powders In Dry Air

Soumitri Seshadri, Mechanical Engineering, College of Staten Island, 60 Farrell Court, Staten Island, NY 10306 and Edward L. Dreizin, Chemical Engineering, New Jersey Institute of Technology, University Heights, Newark, NJ 07102.

Fully dense, nano-composite Al-MoO_3 powders with excess Al were prepared by Arrested Reactive Milling (ARM) as potential components for propellants, explosives, and pyrotechnics. The thermite reaction reducing MoO_3 is expected to boost the Al ignition rate. It is desired to determine the minimum concentration of MoO_3 that enables rapid and complete combustion of aluminum. In the first part of this investigation, nanocomposite powders with the bulk composition $4AlMoO_3$ and $8AlMoO_3$ were produced and their combustion was compared to that of pure aluminum powder in air. The combustion experiments were conducted in a 9.2 L closed vessel. The mass load of aluminum powder was determined from thermodynamic calculations to ensure the maximum flame temperature. To maintain consistency in the experiment, the amount of the nano-composite powder loaded in the vessel was selected to match the volume of the pure Al load. Specifically, 2.89 g, 3.81 g and 3.48 g of Al, $4AlMoO_3$ and $8AlMoO_3$ respectively, were aerosolized in the explosion vessel and ignited at its center using a heated wire while monitoring the vessel pressure in real time as the flame propagated outward. The results showed that both the pressure increase and rate of pressure rise produced in the vessel by combustion of the $AlMoO_3$ mixture is greater than that of Al. Furthermore, the mixture with the lesser concentration of MoO_3 performed better than the mixture with a greater concentration of MoO_3. The project will continue with producing and testing nanocomposite powders with increasingly greater concentrations of Al to establish the minimum concentration of MoO_3 needed to achieve rapid and complete combustion. A special care is taken to ensure the similarity of the particle size distributions for different powders used in the constant volume explosion tests.

454a Desulfurization of Coal Using Microwave Assisted Chemical Treatments

Sapna Jain, Department of Chemical Engineering, Indian Institute of Technology Delhi, Hauz Khas, New Delhi-110016, India and Ratna Choudhury, Centre for Energy Studies, Indian Institute of Technology Delhi, India, New Delhi, India.

India has a multi-fuel scenario in which as much as 46% is covered by non-commercial fuels such as fire wood, animal dung etc., while the commercial fuels like coal, oil and nuclear fuels cover 54 % of the total energy demand. Coal has established itself as dependable, commercial and prime energy source. But its complete utilization cannot be achieved because of the presence of high levels of sulfur. Although Indian coals are low in sulfur but certain reserves in the north-eastern region such as Assam coals, contain high percentage of sulfur. Selective heating by microwave exposure acts as a source of weakening or breaking of bonds. Experiments were conducted to explore the feasibility of desulfurization by microwave radiations. Chemical methods using acid, alkali and solvent vapour extraction of coal along with microwave irradiation were also investigated. Microwave irradiation as pre-treatment method was adopted for simplification of coal structure so that enhanced desulfurization could be achieved. Treatment with microwave followed by solvent vapour extraction resulted in more than 60 % sulfur removal. Speed was the salient feature of desulfurization with microwave treatment. Microwave treatment of alkali-coal, slurry was a one step process, which resulted in significant reduction in total sulfur value and required no other treatment. Chemical and infrared analysis indicated the removal of not only pyritic sulfur but also organic sulfur from Desulfurization by microwave appears to be successful technique, capable of removing all forms of sulfur without affecting coal structure and coking characteristics. Alkali-coal slurry samples on irradiation yielded better results as compared to the dry-coal samples on microwave exposure. The only constraint in adopting microwave-desulfurization process is the economic factor.

REVIEW OF EXPERIMENTAL CAPABILITIES AND HYDRODYNAMIC DATA FOR VALIDATION OF CFD BASED PREDICTIONS FOR SLURRY BUBBLE COLUMN REACTORS

Donna Post Guillen and Daniel S. Wendt
Idaho National Laboratory, Idaho Falls, ID, USA

Steven P. Antal and Michael Z. Podowski
Rensselaer Polytechnic Institute, Troy, NY, USA

Abstract

The purpose of this paper is to document the review of several open-literature sources of both experimental capabilities and published hydrodynamic data to aid in the validation of a Computational Fluid Dynamics (CFD) based model of a slurry bubble column (SBC). The review included searching the Web of Science, ISI Proceedings, and Inspec databases, internet searches as well as other open literature sources. The goal of this study was to identify available experimental facilities and relevant data. Integral (i.e., pertaining to the SBC system), as well as fundamental (i.e., separate effects are considered), data are included in the scope of this effort. The fundamental data is needed to validate the individual mechanistic models or closure laws used in a Computational Multiphase Fluid Dynamics (CMFD) simulation of a SBC. The fundamental data is generally focused on simple geometries (i.e., flow between parallel plates or cylindrical pipes) or custom-designed tests to focus on selected interfacial phenomena. Integral data covers the operation of a SBC as a system with coupled effects. This work highlights selected experimental capabilities and data for the purpose of SBC model validation, and is not meant to be an exhaustive summary.

Introduction

Slurry Bubble Column Reactors (SBCRs) are used by industry to manufacture liquid hydrocarbon fuels (i.e., diesel or gasoline) via the Fischer Tropsch (FT) process (Figure 1). In the FT process, a synthesis gas comprised of hydrogen and carbon monoxide is sparged through a distributor into a suspension of liquid and solid catalyst particles. Generally, in order to be economically viable, a bubble/slurry bubble column reactor must be operated at high volumetric flow rates. This requires a very active catalyst, high catalyst loading of the slurry, and high gas conversion rates. To achieve a complete catalyst suspension and the desired product reaction, these reactors need to be

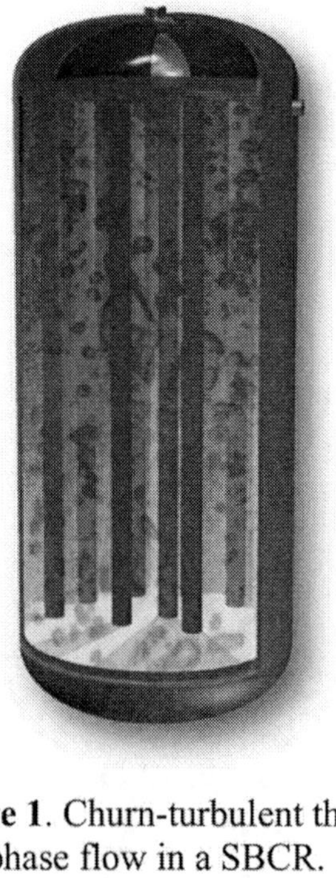

Figure 1. Churn-turbulent three-phase flow in a SBCR.

operated at a high superficial gas velocity in the churn-turbulent flow regime. At these conditions, the two-phase interfacial dynamics dominates the reactor hydrodynamics. Due to the radial gradient of the buoyancy force resulting from non-uniform lateral gas holdup, liquid or slurry recirculation is induced where liquid moves upward in the central region of the bubble column and downward near the wall region of the column. In the case of a SBC, the catalyst particles are very small, hence they closely follow the motion of the liquid flow. The key to successfully modeling this process lies in accurate predictions of the heat and mass transfer along with turbulent mixing, which affects kinetics and thus product yield and selectivity.

Although the FT reaction process was developed in the early 1900's [1], details concerning the hydrodynamic processes, which control the reactor flow, are still poorly understood. For example, significant progress is clearly needed to better understand the unsteady, multiphase fluid dynamics that controls the fluid mixing and interphase transport processes, which in turn determine the overall reactor performance. Due to the complexity of the reaction and hydrodynamic processes occurring in the SBCR, the system performance has traditionally been characterized empirically, rather than from a fundamental physical basis. Since empirical correlations are generally valid only for the (typically narrow) parameter ranges over which they are generated, mechanistic models of the sub-processes occurring in a SBCR as necessary to optimize the overall process and scale laboratory data to industrial applications.

This paper provides a limited summary of institutions with existing experimental facilities capable of providing validation data and sources of validation data published in the open literature. An emphasis is placed on data for churn-turbulent flows, since the hydrodynamics of the flow, as well as the flow mechanisms, change significantly from one flow regime to another [2]. The scope of this effort is limited to hydrodynamics only, without reactions occurring. Once the computational model is validated using suitable hydrodynamic data, reaction kinetics will be incorporated.

Hydrodynamic Data in the Open Literature

The Idaho National Laboratory and Rensselaer Polytechnic Institute (RPI) have embarked on a joint effort to develop scientific and technological advances for the design and development of next generation FT reactors. This goal will be achieved by employing state-of-the-art modeling and computational concepts of multidimensional, multiphase reacting flows in complex geometries and linking together multiple scales to upgrade multi-scale simulation capabilities of the computational multiphase fluid dynamic (CMFD) code, NPHASE [3] under development at RPI. An ensemble-averaged, multifield, mechanistic model formulation will be used to develop closure relations for use with the NPHASE software. Once validated and fully functional, the FT SBCR model will be used as a numerical test bed to virtually simulate new concepts and ideas to improve the process.

The complexity in the design of gas-liquid-solid systems lies in the existence of the three phases with mass, momentum and energy transfer occurring between them. The interfacial structure between the liquid and the gas phase can be considerably different depending upon system configuration and operating conditions. Due to complicated interaction of phases (particularly in churn-turbulent flow regime), the hydrodynamics is not yet fully understood and hence, the reactor design and scale-up are a challenging task. The use of CMFD, rather than empirical correlations, is in principle applicable to a wider range of conditions and design configurations. However, this procedure calls for a solution of the coupled continuity, momentum and energy equations for two fluid phases and a dispersed solid phase. For problems of practical utility, the ensemble-averaged form of the Navier-Stokes equations are solved. This approach necessitates the use of constitutive equations (i.e., closure models) to re-introduce information that was lost through the averaging procedure. Mechanistic-based closure relations have been and are continuing to be developed [4], but require validation with experimental data obtained at conditions applicable to the FT process. To validate a CMFD based model of a SBCR, the important interfacial mass, momentum and energy transfer processes must be identified. For the SBCR under consideration, the transfer processes listed in Table 1 are considered dominant and should be validated with data.

Table 1. Important mass, momentum and energy processes in a SBCR

Mass Transfer	Momentum Transfer	Energy Transfer
• Bubble Coalescence - small bubble interaction - cap/slug bubble interaction • Bubble Breakup - cap/slug bubble breakup • Flow Topology Transition - interfacial area density transport	• Interfacial Drag Force - bubble drag - cap/slug drag - particle drag • Interfacial Lift Force • Interfacial Virtual Mass Force • Interfacial Wall Force • Interfacial Turbulence Dispersion Force • Liquid Turbulence - shear induced turbulence - bubble induced turbulence	• Reaction Kinetics • Tube Bank Heat Transfer • Interfacial Heat Transfer

The purpose of this review is to identify available experimental facilities, techniques and data to validate CMFD models of gas-liquid-solid flows through SBCRs. The following sections will review results from our database and internet searches. In this paper, we refer to "integral" experiments as pertaining to a system, whereas "fundamental" experiments provide the capability to isolate separate effects occurring within a system.

Experimental Facilities, Techniques and Methods Available

The experimental facilities, techniques and methods listed cover both integral and fundamental testing. Capabilities span the scale from bench-top or laboratory SBCs to pilot or production scale slurry bubble column reactors. Available integral test facilities and data available in the open literature are shown in Table 2. Testing performed at these facilities has provided gas, liquid, and solid phase velocity field data; turbulence parameters (turbulent dispersion coefficients, Reynolds stresses); effect of process variables and component properties on gas and solids holdup; bubble size, distribution, and frequency data; as well as heat and mass transfer coefficient data. The data includes bubble column flows between large flat plates (i.e., 2D bubble column experiments) and cylindrical test sections. The 2D bubble column experiments listed were typically smaller in scale than the cylindrical SBCs and range in depth (i.e., distance between the large flat plates) from 5.08 to 15.0 cm and width from 15.0 to 30.48 cm. Although the 2D tests are less expensive to build and operate due to lower flow requirements (which translates to reduced pump and compressor requirements), the effects of the large flat plate walls may significantly influence the two-phase flow phenomena.

The cylindrical bubble columns found in the literature ranged in size from bench-top or laboratory scale to pilot and production size units. The laboratory scale integral tests ranged in diameter from 5.0 to 63.0 cm. The pressure and gas flow rate was typically limited by the material structural integrity and the size of gas compressors used. Therefore, the available data in the churn-flow regime was limited and additional experiments may be needed.

Available open source data (integral and fundamental) is listed in Table 3. The fundamental data is needed to validate individual mechanistic models or closure laws used in a CMFD simulation of a SBC. The fundamental data set is generally taken in more simple geometries (i.e., flow between parallel plates or vertical cylindrical pipes) or custom-designed tests to maximize selected interfacial phenomena. Data available in open literature includes gas, liquid, and solid phase velocity fields; turbulence parameters (liquid phase turbulent dispersion coefficients, Reynolds stress, turbulence intensity); gas and solids holdup distributions; bubble rise velocity, bubble drag coefficient, interfacial area; bubble size distribution and frequency; bubble coalescence and break-up parameters; heat and mass transfer coefficients; and residence time distributions. Test configurations were found to range from simple geometries to more prototypic vessels. The fundamental tests are typically done at low pressure and most times used air-water for the modeling fluids. The 2D test sections used for fundamental testing ranged in depth from 1.2 to 4.0 cm and width from 20 to 30 cm. More recent fundamental testing also includes local velocity and void fraction data in cylindrical vessels and pipes. The diameter for these tests ranged from 3.7 cm to 80 cm. The majority of experiments found in the literature were performed using air-water systems. To validate performance of FT SBCs, it will be necessary to include the effect of operating conditions (e.g., temperature, pressure, etc.) on fluid properties (e.g., density, viscosity, surface tension, etc.) and behavior (e.g., turbulent mixing, gas hold up, etc.).

Summary and Conclusions

This study reviewed existing experimental facilities and available data that can be used to validate a CMFD based model of a SBC. The review included searching the Web of Science, ISI Proceedings, and Inspec databases, internet searches as well as other open literature sources. Experimental facilities, techniques and methods, including bench-top or laboratory scale bubble columns and pilot or production scale slurry bubble column reactors, used to obtain hydrodynamic data relating to SBCs has been summarized. Available experimental data, both integral and fundamental, has also been summarized. Fundamental data is needed to validate individual mechanistic models or closure laws used in a Computational Multiphase Fluid Dynamics (CMFD) simulation of a SBC. The fundamental data set is generally taken in more simple geometries (i.e., flow between parallel plates or cylindrical pipes) or a custom designed tests to maximize selected interfacial phenomena. Integral data can be used to validate the performance of an entire SBC system. The goal of this effort is to produce a validated CMFD based model of a FT SBC that can be used as a numerical test bed. With such a tool, new concepts can be assessed virtually and optimized for FT process improvement.

The authors recognize that an enormous amount of work has been done pertaining to SBCs. The tables provided in this review are meant to be a starting point and are not exhaustive. Future experiments should build upon the work already done and existing data should be used to guide further experimentation.

Table 2. Experimental Capabilities to Validate CMFD-Based Model.

Experimental Facility	Test Section(s)	Measurement Technique	Information Provided	Ref.
Aalborg University Esbjerg, Denmark	0.15 m width × 0.15 m depth × 1 m height	Particle Image Velocimetry (PIV), laser Doppler anemometry (LDA)	gas and liquid phase velocity fields, gas and liquid velocity fluctuations	[5]
Advanced Fuels Development Unit (AFDU), LaPorte, TX	0.46 m dia × 15.24 m height	Radioactive tracer measurements employing NaI scintillation detectors	radioactive tracer response, mean liquid phase axial and radial eddy dispersion coefficient, centerline liquid velocity, mean recirculation velocity	[6]
Beijing Institute of Petrochemical Technology, China (Haibo Jin)	0.3 m dia × 6.6 m height, 0.54 m/s gas velocity, 1.0 MPa system pressure	gamma ray attenuation	effects of superficial gas velocity, static liquid height, liquid surface tension, liquid viscosity, acid concentration, solid and antifoam agents, and system pressure on the axial and overall gas holdup	[7]
Delft University of Technology, The Netherlands (Robert F. Mudde, Wouter K. Harteveld)	15.2 cm, 23.4 cm, 38.4 cm dia columns; 15.0 cm dia × 1.2 m column	LDA, glass fiber probe	axial and tangential liquid velocity components, Reynolds stresses, turbulence power spectra, gas fraction profiles, axial mean liquid velocity, axial normal stress profiles	[8] [9] [10]
Florida Atlantic University (D. Moslemian)		Computer Automated Radioactive Particle Tracking (CARPT)	Mean circulation profiles, turbulence parameters (Reynolds normal and shear stresses, turbulent eddy dispersion coefficient),	[11]
Humboldt University Berlin, Germany (U. Kertzscher)	104 mm dia × 100 mm height SBC	XPTV	local solid velocity and the local solid hold-up in three-phase flows	[12]
Illinois Institute of Technology (Dimitri Gidaspow)	30.48 cm width × 5.08 cm depth × 213.36 height	PIV with γ- and x-ray densitometer	Flow profile, particle concentration profiles and Reynolds stresses PIV – time-averaged particle velocities and concentrations	[13]
Iowa State University (Theodore J. Heindel)	32.1 cm dia × 4.88 m height acrylic SBC	X-ray Computed Tomography (CT)	cross-sectional gas holdup distribution	[14]
NASA Glenn Research Center	25 mm dia × 20 cm height tube	MRI	void fraction distributions	[15]
Sandia National Laboratories (Tim O'Hern, Rob Tachau, John Torczynski, Joel Lash)	Industrial-scale riser testbed; 48 cm dia, ~3 m height, T≤200°C, P≤100 psig	Gamma densitometry tomography, electrical impedance tomography, electrical and optical probes	gas volume fraction, three-phase profiles	[16]
Technische Universität Braunschweig, Germany	0.63 m dia × 6.1 m height Plexiglas bubble column	DPM/TDR, DPM/ECM	local gas and solids holdup	[17]

Experimental Facility	Test Section(s)	Measurement Technique	Information Provided	Ref.
Texas A&M University (Dragomir B. Bukur)	0.05 and 0.21 m inside diameter by 3 m tall stainless steel bubble columns	Gamma ray densitometry	radial gas holdup distribution, radial and axial gas and solid volume fraction distributions, flow regime transitions	[18]
Ohio State University (L. S. Fan)	10 cm dia × 100 cm height bubble column; 10.2 cm width × 10.2 cm depth × 150 cm height Plexiglas square SBC	Electrical Capacitance Tomography, LDV, PIV	real-time cross-sectional gas and solids holdup distributions, instantaneous and average velocity distributions, turbulent energy distributions, Reynolds stresses, liquid-phase power spectra, bubble-induced turbulence, effect of gas distributors and solids on the turbulence field, bubble sizes and distributions	[19] [20] [21] [22] [23]
Purdue University (M. Ishii)	10 mm × 2950 mm high	conductance probe, high speed video, LDA	void fraction, interfacial area concentration, bubble size, gas and liquid velocity	[24]
University of Amsterdam, The Netherlands (R. Krishna)	0.1 m dia, 0.174 m dia, 0.19 m dia, 0.38 m dia, 0.63 m × 4 m height polyacrylate columns	dynamic gas disengagement, pitot tube	total gas holdup; small bubble holdup; dense-phase gas voidage; large bubble (dilute phase) holdup; influence of column diameter, liquid properties, gas distributor, and gas density; rise velocity and average size of large bubbles; centerline liquid velocity; radial distribution of liquid velocity; liquid phase dispersion coefficient	[25] [26]
University of Cambridge, UK (A.J. Sederman)	50 mm dia.	magnetic resonance imaging	gas phase volume fractions, distributions of gas bubble length and velocity	[27]
University of Liege, Belguim (E. Fransolet)	0.24 m dia × 2.75 m height	electrical resistance tomography, wall mounted pressure probes	influence of liquid rheology on gas flow pattern in bubble column reactor, average gas holdup, gas phase distribution, bubble size distribution	[28]
University of Mumbai, India (J. B. Joshi)	385 mm dia × 3.2 m height Perspex bubble column, $u_G = 0.06 - 0.3$ m/s	gamma ray tomography	radial variation of gas holdup	[29]
Washington University (Muthanna Al-Dahhan, Milorad P. Dudukovic)	0.162 m dia × 2.5 m height SBC, P≤1.2 MPa; 30.5 cm dia plexiglass SBC	Computer Automated Radioactive Particle Tracking (CARPT), Computed Tomography (CT), four-point probe, optical oxygen probe	Axial and radial eddy diffusivities, solids residence time, ensemble-average velocities and eddy diffusivities, incipient particle motion, flow regime identification, cross-sectional gas holdup distribution, bubble frequency, specific interfacial area, bubble chord length, bubble velocity, volumetric gas-liquid mass transfer coefficient (k_La), liquid side mass transfer coefficient	[30] [31] [32] [33] [34]

Table 3. Experimental Data to Validate CMFD-Based Model.

System gas/liquid/solid	Column Dimensions	Operating Conditions	Data Acquired	Ref.
air/water	0.15 m dia × 0.66 m height	ambient T and P	bubble pulsation count and frequency, bubble size and distribution	[35]
nitrogen/hydrocarbon waxes/iron oxide and silica particles	0.05 m dia × 3 m height; 0.21 m dia × 3 m height	T = 265°C, atmospheric pressure	radial gas holdup distribution, radial and axial gas and solid volume fraction distributions, flow regime transitions	[18]
nitrogen/FT-300 paraffinic wax	0.051 m dia × 3.05 m height	T = 230-280°C, atmospheric pressure	average gas hold-up, flow regime transition	[36]
Air/water/solids; Air/water, ethanol/solids	30.5 cm dia × 100 cm height		cross sectional gas and solids holdup distribution, time-averaged particle velocities	[30]
dimethyl ether synthesis SBC reactor	0.46 m dia × 15.24 m height	T = 250°C, P = 5.27 MPa	radioactive tracer response, mean liquid phase axial and radial eddy dispersion coefficient, centerline liquid velocity, mean recirculation velocity	[6]
air/water	200, 400, 800 mm dia × 3 m height	u_G = 20-90 mm/s	instantaneous local heat transfer rates	[37]
nitrogen, carbon dioxide/ cyclohexane, water	0.2 m dia × 1.6 m height	T = ~20°C, P_{atm}, $u_G \leq 0.14$ m/s	mass transfer efficiency, volumetric mass transfer coefficient (k_La)	[38]
air/water/500 μm acetate particles	10.2 cm width × 10.2 cm depth × 150 cm height	u_G = 0.025-7.5 cm/s	turbulent energy distributions, average velocity and Reynolds stresses, liquid-phase power spectra, bubble-induced turbulence, effect of gas distributors on the turbulence field, effect of solids on the liquid-phase turbulence	[39] [22]
air/water	14 cm dia; 19 cm dia; 44 cm dia	u_G = 2.0-12.0 cm/s	time-averaged spatial flow structure, axial liquid velocity profiles, Reynolds shear stress, and turbulence intensities profiles	[40]
air/water, K_2SO_4/poly-methyl methacrylate, polyoxymethylene	0.63 m dia × 6.1 m height Plexiglas bubble column	u_G = 0.02-0.09 m/s	local gas and solids holdup	[17]
air/various Newtonian and non-Newtonian fluids	0.24 m dia × 1.60 m height; 0.30 m dia × 0.50 m height	T = 293K	bubble rise velocity, drag coefficient	[41]
air/aqueous xanthan solutions	0.24 m dia × 2.75 m height	u_G = 0.02-0.15 m/s	influence of liquid rheology on gas flow pattern, average gas holdup, gas phase distribution, bubble size distribution	[28]
air/water	0.162 m dia	$P \leq 1.0$ MPa	overall gas holdup, volumetric gas-liquid mass transfer coefficient (k_La), interfacial area, liquid side mass transfer coefficient	[34]
air/water	0.4 m dia × 9 m height	atmospheric pressure	gas-liquid mass transfer, total gas hold-up, small bubble hold-up in the liquid–small bubble mixture, large and small bubble hold-up	[42]
H_2, N_2, CO, CH_4/hexane mixture/FT Fe catalyst	0.316 m dia × 2.8 m height	$P \leq 0.8$ MPa	effects of gas velocity, system pressure, and catalyst loading on gas holdup	[43]
air/water, Drakeol 10	48 cm dia × ~3 m height	$P \leq 50$ psig $u_G \leq 0.25$ m/s	spatially resolved gas holdup	[16]
air/water; air/acetic acid	0.3 m dia × 6.6 m height	$P \leq 1.0$ MPa $u_G \leq 0.4$ m/s	effects of superficial gas velocity, liquid surface tension, liquid viscosity, and system pressure on the axial gas holdup	[7]

System gas/liquid/solid	Column Dimensions	Operating Conditions	Data Acquired	Ref.
nitrogen, helium/ethanol, 1-butanol, toluene, decalin, tap water	0.1 m dia × 2.4 m height	T = 293K, P = 0.1-4.0 MPa, u_G = 0.01-0.20 m/s	effect of gas density on the gas hold-up structure in a bubble column with organic liquids	[44]
nitrogen/distilled water, sodium gluconate/carbon particles	0.3 m width × 0.015 m depth × 2 m height	T = 293 K, P = 1 bar	local and overall gas holdup in homogenous, transition, and heterogeneous regimes	[45]
air/water	0.174, 0.38, 0.63 m dia × 4 m height	P = 101.3 kPa	radial distribution of liquid velocities, liquid-phase axial dispersion coefficients, bubble drag coefficient	[46]
air/paraffin oil/porous silica particles	0.1, 0.19, 0.38 m dia	P = 1 bar	gas holdup, gas voidage, rise velocity and average size of large bubbles, centerline liquid velocity, radial distribution of liquid velocity, liquid phase dispersion coefficient	[26]
air/water, isopropanol	0.10, 0.14, 0.19, 0.26, 0.30 m dia	u_G = 0.02-0.12 m/s	effect of column diameter, superficial gas velocity, distributor type, static liquid height, axial distance from distributor, and liquid properties on gas-holdup	[47]
air/water, inorganic industrial solution	0.078 m dia × 4.6 m height	T = 45°C, u_G = 1-8 cm/s	average gas holdup, bubble size distribution, method to estimate coalescence and break-up parameters in bubble columns	[48]
air/1.5 wt% polyacryamide in water	30 cm width × 1.2 cm depth × 120 cm height		coalescing mechanism of two in-line oblate-cusped bubbles rising in a non-Newtonian fluid	[49]
air/water, glycerin	68 mm width × 88 mm depth × 450 mm height		turbulence intensity, Reynolds stress, liquid velocity profile associated with bubble induced flow structure	[50]
nitrogen/water, ethanol, 1-butanol	0.1 m dia × 2.1 m height	T = 298-323K, P = 0.1-0.5 MPa, u_G = 0.01-0.21 m/s	gas holdup, axial liquid velocity, liquid axial dispersion coefficient, CFD model developed for the prediction of flow pattern in terms of mean velocity and eddy diffusivity profiles. Model also predicts residence time distribution and axial dispersion coefficient.	[51]
air/water/calcium alginate beads	0.14 m dia		effect of presence of solid phase on homogeneous–heterogeneous flow regime transition, regime transition critical point, voidage as function of gas flow rate	[52]
air/water, n-butanol, glycerin	10 cm width × 10 cm depth × 150 cm height	ambient T and P	effect of liquid properties on average gas holdup values, bubble size distributions and Sauter diameters	[53]
air/water	0.2 m width × 0.04 m depth × 1.2 m height	ambient T and P	flow regime identification, Reynolds stresses, global gas hold-up, liquid velocities, liquid flow macrostructures	[54]
coal-to-liquids reaction	1 m dia × 11.8 m height pilot plant	T = 313-733 K, P = 16.6-16.8 MPa	residence time distribution curves	[55]
air/water, cationic and anionic surfactant aqueous solutions	0.05 m dia × 0.40 m height glass bubble column	T = 20°C, dynamic bubble regime	effect of liquid properties (surfactants) on bubble generation phenomenon, interfacial area, and liquid-side mass transfer coefficient	[56]
air/clay particles suspended in water	0.3 m dia × 3 m height		effect of power input, fluid phase viscosity and solids loading on the mechanical stress on suspended particles	[57]

System gas/liquid/solid	Column Dimensions	Operating Conditions	Data Acquired	Ref.
nitrogen/cyclohexane	0.3 m dia × 4 m height	T = 30-160°C P = 0.2-1.1 MPa	bubble size distribution, mean gas holdup, interfacial area, bubble swarm velocity	[58]
air/water; steam/water	195.3 mm dia	20° C < T < 30° C P = 120 kPa; T=180-280° C 1 < P < 6.5 MPa	gas fraction, gas velocity, bubble size	[59]
air/water/glass beads (150 μm)	0.162 m dia × 2.5 m height	P = 0.1-1.0 MPa u_G = 0.08-0.45 m/s	effect of reactor pressure and superficial gas velocity on solids phase velocity and shear stress	[32]
air/water/glass beads (150 μm)	0.162 m dia × 2.5 m height	P = 0.1-1.0 MPa u_G = 0.08-0.45 m/s	cross sectional holdup distribution of gas, liquid, and solid phases	[60]
air/water	0.1 m width × 0.1 m depth × 1.0 m height	u_G = 0-8 cm/s	bubble velocity, diameter, void fraction, and liquid velocity, influence of the void fraction on the relative velocity of a swarm of gas bubbles, experimental drag correlation proposed	[61]
nitrogen/water/glass beads	8.89 cm dia × 290 cm height	u_G = 0.5-12 cm/s, 5-30 wt% solids	solid holdup	[62]
nitrogen/Tellus oil, aqueous glucose solutions	0.15 m dia × 1.22 m height; 0.23 m dia × 1.22 m height	ambient temperature, P = 0.1-1 MPa	effect of high liquid viscosity, column diameter and operating pressure on the total gas hold-up	[63]
air, nitrogen/water, organic oil/activated carbon, silica particles	0.15 m dia	ambient temperature, P = 0.1-1.3 MPa	influence of particle lyophobicity on gas hold-up, homogeneous to churn-turbulent regime transition, and gas–liquid mass transfer	[64]
air/C_9-C_{11} paraffin oil/porous catalyst	0.1 m dia	ambient T and P, u_G = 0-0.4 m/s	gas holdup and volumetric mass transfer coefficient in homogeneous and churn-turbulent flow regimes	[65]
air/aqueous solutions of n-butanol	0.385 m dia × 3.2 m height column	u_G = 0.06-0.24 m/s	effect of superficial gas velocity on radial gas hold-up profiles	[29] [66]
air/water	0.1 m dia × 2 m height column	ambient T and P, u_G = 0.6-15 cm/s	axial and tangential mean and RMS liquid velocity values	[67]
air/water	0.16 m dia × 2.5 m height column	P ≤ 10 bar u_G = 0.03-0.30 m/s	time-averaged local heat transfer coefficients	[68]
water, hydrocarbon liquids	5.08 cm dia × 55.88 cm height, 10.16 cm dia × 91.44 cm height columns	room temperature, P ≤ 10.3 MPa u_G ≤ 0.4 cm/s	liquid phase axial dispersion coefficients	[69]
nitrogen, carbon monoxide, hydrogen/liquid paraffin/silica gel powder	37 mm dia × 480 mm height SBC	T = RT-300°C, P = atm-6 MPa u_G ≤ 0.02 m/s	bubble sizes, interfacial areas and volumetric mass transfer coefficients	[70]
air/water	10 mm thick × 2950 mm high	T = 70 °C, P = 24-49 kPa	void fraction, interfacial area concentration, interaction mechanisms	[24]
air/water	0.114, 0.190, 0.292 m dia columns	u_G = 2-18.4 cm/s	axial dispersion coefficients (eddy diffusivities)	[71]

References

1. Fischer, F. and H. Tropsch, *Synthesis of Methanol for CO and H2*, French Patent 540, Editor. 1921.
2. Chen, Y., *Modeling Gas-Liquid Flow in Pipes: Flow Regime Transitions and Drift-Flux Modeling.* 2001, Stanford University.
3. Antal, S.P., et al., *Development of a Next Generation Computer Code for the Prediction of Multicomponent Multiphase Flows*, in *International Meeting on Trends in Numerical and Physical Modeling for Industrial Multiphase Flow*. 2000: Cargese, France.
4. Podowski, M.Z., *On the Mechanistic Modeling of Interfacial Phenomena in Gas/Liquid Two-Phase Flows*, in *6th International Conference on Multiphase Flows, ICMF 2007*. 2007: Leipzig, Germany.
5. Deen, N.G., B.H. Hjertager, and T. Solberg, *Comparison of PIV and LDA Measurement Methods Applied to the Gas-Liquid Flow in a Bubble Column*, in *10th International Symposium on Applications of Laser Techniques to Fluid Mechanics*. 2000: Lisbon, Portugal.
6. Chen, P., et al., *Hydrodynamics of slurry bubble column during dimethyl ether (DME) synthesis: Gas-liquid recirculation model and radioactive tracer studies.* Chemical Engineering Science, 2006. **61**(19): p. 6553-6570.
7. Jin, H.B., et al., *The axial distribution of holdups in an industrial-scale bubble column with evaluated pressure using gamma-ray attenuation approach.* Chemical Engineering Journal, 2005. **115**(1-2): p. 45-50.
8. Mudde, R.F., J.S. Groen, and H.E.A.V.D. Akker, *Liquid velocity field in a bubble column: LDA experiments.* Chemical Engineering Science, 1997. **52**(21-22): p. 4217-4224.
9. Harteveld, W.K., R.F. Mudde, and H.E.A. Van den Akker, *Estimation of turbulence power spectra for bubbly flows from laser Doppler Anemometry signals.* Chemical Engineering Science, 2005. **60**(22): p. 6160-6168.
10. Harteveld, W.K., R.F. Mudde, and H.E.A. van den Akker, *Dynamics of a bubble column: Influence of gas distribution on coherent structures.* Canadian Journal of Chemical Engineering, 2003. **81**(3-4): p. 389-394.
11. Moslemian, D., N. Devanathan, and M.P. Dudukovic, *Radioactive particle tracking technique for investigation of phase recirculation and turbulence in multiphase systems.* Review of Scientific Instruments, 1992. **63**(10): p. 4361-4372.
12. Seeger, A., et al., *Measurement of the local velocity of the solid phase and the local solid hold-up in a three-phase flow by X-ray based particle tracking velocimetry (XPTV).* Chemical Engineering Science, 2003. **58**(9): p. 1721-1729.
13. Lam, P. and D. Gidaspow, *Computational and Experimental Modeling of Slurry Bubble Column Reactors.* 2000, University of Akron, Illinois Institute of Technology.
14. Hubers, J.L., et al., *X-ray computed tomography in large bubble columns.* Chemical Engineering Science, 2005. **60**(22): p. 6124-6133.
15. Daidzic, N.E., et al., *Gas-liquid phase distribution and void fraction measurements using MRI.* Nuclear Engineering and Design, 2005. **235**(10-12): p. 1163-1178.
16. Jackson, N.B., et al., *Sandia Support for PETC Fischer-Tropsch Research: Experimental Characterization of Slurry-Phase Bubble-Column Reactor Hydrodynamics.* 1996, Advanced Energy Technology Center, Engineering Sciences Center.
17. Dziallas, H., V. Michele, and D.C. Hempel, *Measurement of Local Phase Holdups in a Two- and Three-Phase Bubble Column.* Chemical Engineering & Technology, 2000. **23**(10): p. 877-884.

18. Bukur, D.B., J.G. Daly, and S.A. Patel, *Application of gamma-ray attenuation for measurement of gas holdups and flow regime transitions in bubble columns.* Industrial & Engineering Chemistry Research, 1996. **35**(1): p. 70-80.
19. Warsito, W. and L.-S. Fan, *Measurement of real-time flow structures in gas–liquid and gas–liquid–solid flow systems using electrical capacitance tomography (ECT).* Chemical Engineering Science, 2001. **56**(21-22): p. 6455-6462.
20. Warsito, W. and L.S. Fan, *3D-ECT velocimetry for flow structure quantification of gas-liquid-solid fluidized beds.* Canadian Journal of Chemical Engineering, 2003. **81**(3-4): p. 875-884.
21. Warsito, W. and L.S. Fan, *Dynamics of spiral bubble plume motion in the entrance region of bubble columns and three-phase fluidized beds using 3D ECT.* Chemical Engineering Science, 2005. **60**(22): p. 6073-6084.
22. Cui, Z. and L.S. Fan, *Energy spectra for interactive turbulence fields in a bubble column.* Industrial & Engineering Chemistry Research, 2005. **44**(5): p. 1150-1159.
23. Chen, R.C. and L.-S. Fan, *Particle Image Velocimetry for Characterizing the Flow Structure in Three-Dimensional Gas-Liquid-Solid Fluidized Beds.* Chemical Engineering Science, 1992. **47**(13/14): p. 3615-3622.
24. Sun, X., *Two-Group Interfacial Area Transport Equation for a Confined Test Section.* 2001, Purdue University.
25. Krishna, R. and J. Ellenberger, *Gas holdup in bubble column reactors operating in the churn-turbulent flow regime.* Aiche Journal, 1996. **42**(9): p. 2627-2634.
26. Krishna, R., et al., *Design and scale up of a bubble column slurry reactor for Fischer-Tropsch synthesis.* Chemical Engineering Science, 2001. **56**(2): p. 537-545.
27. Sederman, A.J., M.D. Mantle, and L.F. Gladden, *Quantitative 'Real-Time' Imaging of Multi-Phase Flow in Ceramic Monoliths.* Magnetic Resonance Imaging, 2003. **21**: p. 359-361.
28. Fransolet, E., et al., *Analysis of gas holdup in bubble columns with non-Newtonian fluid using electrical resistance tomography and dynamic gas disengagement technique.* Chemical Engineering Science, 2005. **60**(22): p. 6118-6123.
29. Veera, U.P., K.L. Kataria, and J.B. Joshi, *Gas hold-up profiles in foaming liquids in bubble columns.* Chemical Engineering Journal, 2001. **84**(3): p. 247-256.
30. Chen, J., et al., *Particle Motion in Packed/Ebullated Beds by CT and CARPT.* Aiche Journal, 2001. **47**(5): p. 994-1004.
31. Nedeltchev, S., S.B. Kumar, and M.P. Dudukovic, *Flow regime identification in a bubble column based on both Kolmogorov entropy and quality of mixedness derived from CARPT data.* Canadian Journal of Chemical Engineering, 2003. **81**(3-4): p. 367-374.
32. Rados, N., A. Shaikh, and M.H. Al-Dahhan, *Solids flow mapping in a high pressure slurry bubble column.* Chemical Engineering Science, 2005. **60**(22): p. 6067-6072.
33. Xue, J.L., et al., *Bubble dynamics measurements using four-point optical probe.* Canadian Journal of Chemical Engineering, 2003. **81**(3-4): p. 375-381.
34. Han, L. and M.H. Al-Dahhan, *Gas-liquid mass transfer in a high pressure bubble column reactor with different sparger designs.* Chemical Engineering Science, 2007. **62**(1-2): p. 131-139.
35. Al-Masry, W.A., E.M. Ali, and Y.M. Aqeel, *Determination of bubble characteristics in bubble columns using statistical analysis of acoustic sound measurements.* Chemical Engineering Research & Design, 2005. **83**(A10): p. 1196-1207.
36. Bukur, D.B., D. Petrovic, and J.G. Daly, *Flow Regime Transitions in a Bubble Column with a Parrafin Wax as the Liquid Medium.* Industrial & Engineering Chemistry Research, 1987. **26**: p. 1087-1092.

37. Chen, W., et al., *Generalized dynamic modeling of local heat transfer in bubble columns.* Chemical Engineering Journal, 2003. **96**(1-3): p. 37-44.
38. Chaumat, H., et al., *Mass transfer in bubble column for industrial conditions-effects of organic medium, gas and liquid flow rates and column design.* Chemical Engineering Science, 2005. **60**(22): p. 5930-5936.
39. Cui, Z. and L.S. Fan, *Turbulence energy distributions in bubbling gas-liquid and gas-liquid-solid flow systems.* Chemical Engineering Science, 2004. **59**(8-9): p. 1755-1766.
40. Degaleesan, S., M. Dudukovic, and Y. Pan, *Experimental Study of Gas-Induced Liquid-Flow Structures in Bubble Columns.* Aiche Journal, 2001. **447**(9): p. 1913-1931.
41. Frank, X., et al., *Bubble motion in non-Newtonian fluids and suspensions.* Canadian Journal of Chemical Engineering, 2003. **81**(3-4): p. 483-490.
42. Haut, B. and T. Cartage, *Mathematical modeling of gas-liquid mass transfer rate in bubble columns operated in the heterogeneous regime.* Chemical Engineering Science, 2005. **60**(22): p. 5937-5944.
43. Inga, J.R. and B.I. Morsi, *Effect of operating variables on the gas holdup in a large-scale slurry bubble column reactor operating with an organic liquid mixture.* Industrial & Engineering Chemistry Research, 1999. **38**(3): p. 928-937.
44. Jordan, U., A.K. Saxena, and A. Schumpe, *Dynamic gas disengagement in a high-pressure bubble column.* Canadian Journal of Chemical Engineering, 2003. **81**(3-4): p. 491-498.
45. Kluytmans, J.H.J., et al., *2D slurry bubble column hydrodynamic phenomena clarified with a 3D gas-liquid model.* Canadian Journal of Chemical Engineering, 2003. **81**(3-4): p. 456-464.
46. Krishna, R., et al., *Influence of scale on the hydrodynamics of bubble columns operating in the churn-turbulent regime: experiments vs. Eulerian simulations.* Chemical Engineering Science, 1999. **54**(21): p. 4903-4911.
47. Kumar, S.B., D. Moslemian, and M.P. Dudukovic, *Gas-holdup measurements in bubble columns using computed tomography.* Aiche Journal, 1997. **43**(6): p. 1414-1425.
48. Laari, A. and K. Turunen, *Experimental determination of bubble coalescence and break-up rates in a bubble column reactor.* Canadian Journal of Chemical Engineering, 2003. **81**(3-4): p. 395-401.
49. Lin, T.J. and G.M. Lin, *The mechanisms of bubble coalescence in a non-Newtonian fluid.* Canadian Journal of Chemical Engineering, 2003. **81**(3-4): p. 476-482.
50. Liu, Z., et al., *Study of bubble induced flow structure using PIV.* Chemical Engineering Science, 2005. **60**(13): p. 3537-3552.
51. Lorenz, O., et al., *Liquid phase axial mixing in bubble columns operated at high pressures.* Chemical Engineering Science, 2005. **60**(13): p. 3573-3586.
52. Mena, P.C., et al., *Effect of solids on homogeneous-heterogeneous flow regime transition in bubble columns.* Chemical Engineering Science, 2005. **60**(22): p. 6013-6026.
53. Mouza, A.A., G.K. Dalakoglou, and S.V. Paras, *Effect of liquid properties on the performance of bubble column reactors with fine pore spargers.* Chemical Engineering Science, 2005. **60**(5): p. 1465-1475.
54. Olmos, E., C. Gentric, and N. Midoux, *Identification of flow regimes in a flat gas-liquid bubble column via wavelet transform.* Canadian Journal of Chemical Engineering, 2003. **81**(3-4): p. 382-388.
55. Onozaki, M., et al., *Dynamic simulation of gas–liquid dispersion behavior in coal liquefaction reactors.* Chemical Engineering Science, 2000. **55**(21): p. 5099-5113.
56. Painmanakul, P., et al., *Effect of surfactants on liquid-side mass transfer coefficients.* Chemical Engineering Science, 2005. **60**(22): p. 6480-6491.

57. Pilz, R.D. and D.C. Hempel, *Mechanical stress on suspended particles in two- and three-phase airlift loop reactors and bubble columns.* Chemical Engineering Science, 2005. **60**(22): p. 6004-6012.
58. Pohorecki, R., et al., *Hydrodynamics of a pilot plant bubble column under elevated temperature and pressure.* Chemical Engineering Science, 2001. **56**(3): p. 1167-1174.
59. Prasser, H.-M., et al., *Evolution of the structure of a gas-liquid two-phase flow in a large vertical pipe.* Nuclear Engineering and Design, 2007: p. doi:10.1016/j.nucengdes.2007.02.018.
60. Rados, N., A. Shaikh, and M. Al-Dahhan, *Phase distribution in a high pressure slurry bubble column via a single source computed tomography.* Canadian Journal of Chemical Engineering, 2005. **83**(1): p. 104-112.
61. Simonnet, M., et al., *Experimental determination of the drag coefficient in a swarm of bubbles.* Chemical Engineering Science, 2007. **62**(3): p. 858-866.
62. Soong, Y., et al., *Ultrasonic Characterizations of Slurries in a Bubble Column Reactor.* Industrial & Engineering Chemistry Research, 1999. **38**: p. 2137-2143.
63. Urseanu, M.I., et al., *Influence of operating pressure on the gas hold-up in bubble columns for high viscous media.* Chemical Engineering Science, 2003. **58**(3-6): p. 697-704.
64. van der Schaaf, J., et al., *Effect of particle lyophobicity in slurry bubble columns at elevated pressures.* Chemical Engineering Science, 2006.
65. Vandu, C.O., K. Koop, and R. Krishna, *Volumetric mass transfer coefficient in a slurry bubble column operating in the heterogeneous flow regime.* Chemical Engineering Science, 2004. **59**(22-23): p. 5417-5423.
66. Veera, U.P., K.L. Kataria, and J.B. Joshi, *Effect of superficial gas velocity on gas hold-up profiles in foaming liquids in bubble column reactors.* Chemical Engineering Journal, 2004. **99**(1): p. 53-58.
67. Vial, C., et al., *Influence of gas distribution and regime transitions on liquid velocity and turbulence in a 3-D bubble column.* Chemical Engineering Science, 2001. **56**(3): p. 1085-1093.
68. Wu, C., M.H. Al-Dahhan, and A. Prakash, *Heat transfer coefficients in a high-pressure bubble column.* Chemical Engineering Science, 2007. **62**(1-2): p. 140-147.
69. Yang, G.Q. and L.S. Fan, *Axial liquid mixing in high-pressure bubble columns.* Aiche Journal, 2003. **49**(8): p. 1995-2008.
70. Yang, W., et al., *Gas-liquid mass transfer in slurry bubble systems. II. Verification and simulation of the model based on the single bubble mechanism.* Chemical Engineering Journal, 2003. **96**(1-3): p. 29-35.
71. Yang, Y.B., N. Devanathan, and M.P. Dudukovic, *Liquid Backmixing in Bubble Columns.* Chemical Engineering Science, 1992. **47**(9-11): p. 2859-2864.

Potential of Water Hyacinth for Biomass Refining

Lealon Martin, Jessica E. Hronich, and Henry R. Bungay

Howard P. Isermann Department of Chemical Engineering
Rensselaer Polytechnic Institute
Troy, NY 12180-3590 e-mail: lealon@rpi.edu

In this talk, we focus on the sustainable conversion of biomass from aquatic vegetation (featuring Eichhornia crassipes, commonly known as Water Hyacinth, as a model system) into liquid fuel (cellulosic ethanol). From an ecological standpoint, *E. crassipes* exemplifies a fast-growing, nutrient competitive, eco-resilient highly invasive aquatic nuisance. However, from a chemical process standpoint, these same properties (rarely encountered in other plant systems, i.e. corn) are characteristics of an ideal feedstock for renewable biomass. Furthermore, the high water content (> 95 wt%) and low lignin content (~13 %) of *E. crassipes* make it a natural substrate for rapid microbial metabolism that can be employed, partially, as a potentially effective biological pretreatment technology. In comparison to other conventional biomass pretreatment processes (e.g., grinding and milling of corn or steam explosion of wood pulp), this technology has the potential to be more easily implemented, less energy intensive, and relatively inexpensive. We have conducted an economic life cycle analysis to determine whether the combined growth and digestion properties of Water Hyacinth make it competitive and cost-effective source of biomass in comparison to conventional feedstocks. Finally, a sensitivity analysis is performed to identify ecological, biological and chemical processing parameters that could significantly impact the economic viability of *E. crassipes* conversion to biomass.

Introduction

In the United States, some 95% of the ethanol commercially produced for fuel applications comes from corn. The remainder comes from grain sources such as wheat or barley (1). Much work has been completed recently using corn as an ethanol feedstock in the United States. However, according to the DOE, if all the corn grain grown in the United States were converted into ethanol, it would only fulfill approximately 15% of current transportation needs (66). Additionally, subsidies are required to make corn an economically competitive crop. In locations where it can grow (such as Brazil) sugarcane has proven to be an exceedingly effective feedstock for bioethanol production (3). Unfortunately, particularly within the U.S., there are serious concerns regarding the expansion of ethanol production from plant starch. Chief among these is land use --- there may simply not be sufficient cultivatable land to support a significant expansion in ethanol capacity. This is compounded by the fact that most of the plant crops attractive for bioethanol are already used as primary dietary staples (4). In order to meet the Department of Energy goal of 60 billion gallons of ethanol by 2030, a major shift in paradigm from starch-based ethanol production is needed. In response, other feedstocks, such as cellulosic biomass, must be investigated for ethanol production. Cellulosic biomass is derived from vegetation other than food crops, and can be a plentiful and viable source for producing fuel grade ethanol. Wood is considered to be a primary cellulosic biomass feedstock; however it has an extremely high lignin content, and thus requires expensive and energy-intensive pre-treatment, such as steam or ammonia explosion. Switch grasses have also been in investigated as a potential cellulosic feedstock; however, they are reported to have low cellulose yields per cultivated area compared to other feedstocks.

Alternative sources of cellulose could be derived from aquatic vegetation, such as algae. Although they have received considerably less attention, aquatic plants do not require land to grow and thus do not compete with agricultural crops. Also, the harvest frequency for aquatics tends to be on the order of days, whereas the frequency for trees, grasses, and crops are on the order of years and months. While algae does indeed have a high growth rate, it is difficult to grow algae at maximum rates outside of laboratory conditions and it can be extremely difficult to harvest.

Nuisance aquatic plants, such as the water hyacinth, are one such feedstock that have rapid growth rates, low cost, low lignin content, and plentiful availability. Additionally, in areas currently infested with water hyacinths, millions of dollars are spent each year to control the population through harvesting and disposal. Developing a process to produce a valuable energy source from what was once considered a blight on an ecosystem will greatly help the economy of surrounding communities. The plentiful availability, low cost, and rapid growth of water hyacinths make them an ideal candidate for a biofuel, particularly in developing countries. In addition, our preliminary analysis has shown using water hyacinth as a feedstock could economically produce biomass at approximately $44 per ton dry mass, without the benefit of subsidies. This makes biomass feedstock produced from water hyacinth a viable choice, as it can be produced for less than $50 per ton of dry biomass (5). However, since *E. crassipes* has very little native starch content, it will be critical to this research to identify an effective and practical solution for the digestion of cellulosic plant matter.

Manufacturing Cost Estimation

A comprehensive plant design spreadsheet was created with the goal of estimating the cost to convert water hyacinths into a viable biomass feedstock. Values from literature were obtained for variables such as yearly plant growth per acre, plant density, and mat connectivity. Vendors were researched or contacted for energy requirements of commercially available harvesters and roll presses. In an effort to legitimize the estimations made, a standardized manufacturing cost summary sheet for the chemical production industry was adapted for our agribusiness plan (6).

A lake covered with 300 acres of water hyacinth was chosen as the basis for the study. A somewhat conservative annual growth rate of 100 tons DM / ha / yr was used[1]. With an average plant growth density of 14 kg / m^2, (7) this allows for approximately 780 tons of (wet) hyacinth plants to be harvested in a day. This gives the plants a 22 day harvest frequency, which is sufficient for re-growth (8).

As the exact energy requirements of the proposed harvester are not yet known, the 100 kW commercially available Water Witch harvester (9) was used in the estimation. It is expected that the new design will require less energy, and therefore this was used as a maximum value. Assuming an 8 hour work day, 3 harvesters as well as 3 boats to tow the mats to shore will be required to maintain the daily harvesting goal. Scenarios were also run using both barges and a

[1] Under the right conditions, this number could be as high as 150 tons DM / ha / yr. (71)

winch system to transport the floating hyacinth mats to the shoreline for processing; however, boats with a tow line were the most economically feasible.

Once on shore, the hyacinth mats will go through a series of roll presses to remove 97% of the water. This is done both as a pretreatment step, and to reduce the volume for silage. The hyacinth will still contain approximately 36% water by mass after the pressing. The plants will then be placed in silage for approximately 14 days to allow for decomposition of the cell walls. Any supernatant liquid will be decanted, processed, and returned to the lake.

Operators were assumed to work 8 hour shifts, 310 days a year. A base hourly compensation of $10 / hr was used; benefits were accounted for by multiplying operator pay by a factor of 1.3.

The state of Florida spends approximately $4 million annually[2] to control 33,000 acres of water hyacinth (10). Therefore, an annual removal credit of $130 per acre was applied to the cost estimation. The process as described will produce approximately 12,000 tons of dry biomass per year, at a cost of $44 per dry ton. According to previous estimates by Turhollow (11) and the DOE (5), this places water hyacinths within the realm of economically competitive biomass feedstocks. Table I illustrates the contribution of each step into the final cost of the biomass.

Table I: Operating costs per ton of dry biomass.

Operating Costs ($ / dry ton biomass)	
Manufacturing Expenses	
Direct	
Removal Credit	($3.22)
Operating labor	$18.62
Supervisory and clerical labor	$1.86
Utilities	
Electricity	$4.03
Gasoline	$5.73
Maintenance and repairs	$1.83
Operating supplies	$0.18
Indirect	
Overhead, packaging, storage	$5.58
Local taxes	$1.01
Insurance	$2.02
General Expenses	
Administrative costs	$1.39
Depreciation	$5.04
Total cost to produce, per dry ton	$44.08

[2] According to Simberloff et. al, an average annual cost of $2.7 million was spent to manage 13,400 ha of water hyacinth mixed with water lettuce. The cost was adjusted for inflation using the CPI inflation calculator. (11)

Results

Sensitivity Analysis

A sensitivity analysis was performed using the manufacturing cost estimation generated for this process. While all input factors for the process were varied to determine the effect on the final feedstock price per ton, only the most relevant will be presented.

As explained in the process development section, a novel method will be created to efficiently slice the thick mats of hyacinth and then tow them to shore. The width of the hyacinth swath has a pronounced effect on the ultimate price of the biomass, as shown in Figure 2. This effect is due to the number of harvesters (and thus operators) required in a given day. The wider the cut width of the harvesters, the fewer harvesters will be needed to harvest the 780 tons of hyacinth. A maximum width of 7.5 m is reported – wider than this, no further effect on the price is observed. Unfortunately, a width of 7.5 m could prove to add complications in the water removal step. A rolling press and conveyor belt of this length would be required, and the press alone would most likely prove to be infeasible. Therefore, shorter cut lengths will most likely be used. This barrier could perhaps be overcome by either folding the mats of hyacinth before feeding them to the press, or possibly by further cutting the mats on land before pressing.

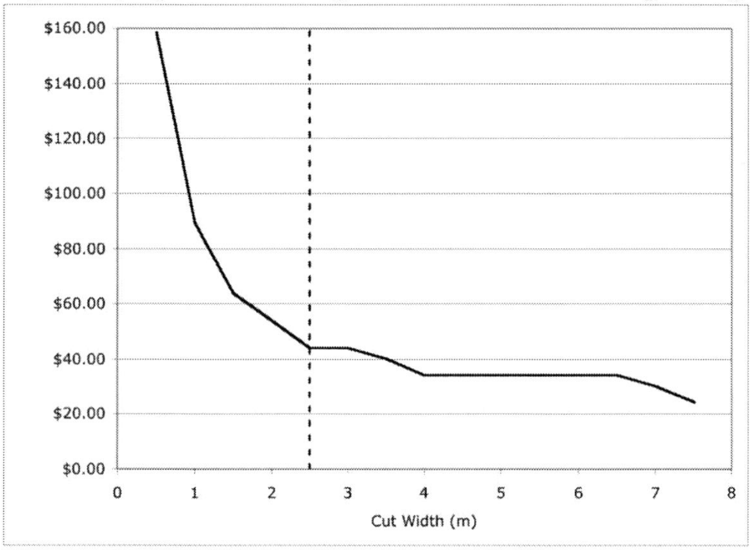

Figure 2: Biomass price per dry ton as a function of harvester cut width in meters. The dotted line represents the width used in the cost estimation.

The speed at which the harvester cuts the mats of hyacinths is another factor affecting final biomass price. As there is no harvester yet designed which would cut the mats as described in the process development section, the speed at which the harvester would travel is somewhat speculation. Other commercially available harvesters can travel at top speeds of 6.5 to 8.5 knots (9), or approximately 200 – 260 m / min. At speeds above 50 m / min, the price no longer drops in our estimation, as shown in Figure 3. Therefore, if the harvester can indeed travel at speeds faster than 50 m / min, the price could drop to $32 per ton of dry biomass. Only one harvester would be required to slice the mats. Depending on the productivity of the day, the harvester operator could then help out with other functions, thus eliminating the need for more operators.

A potential drawback of higher harvester boat speeds could be accelerated degradation of the cutting saw. The optimal speed would have to be discussed with the design manufacturer.

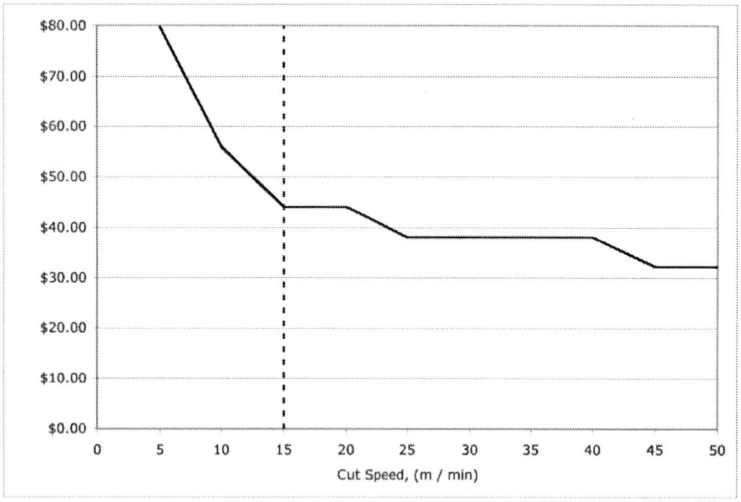

Figure 3: Biomass price per dry ton as a function of harvester cut speed in meters per minute. The dotted line represents the speed used in the cost estimation.

As would be expected, the price of biomass per dry ton increases linearly with both the harvester and transport boat energy requirements. However, the effect is slight, especially in the case of the transport boat. Therefore, if the energy requirements used in the estimation are incorrect, they will have little effect on the overall price of the biomass, as can be seen in Figures 4a and 4b.

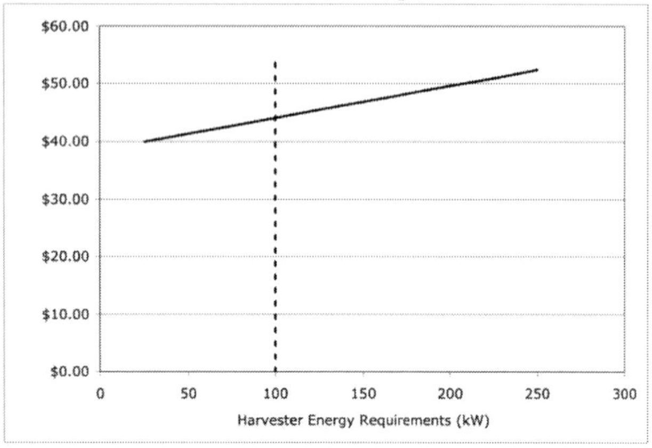

Figure 4a: Biomass price per dry ton as a function of harvester energy requirement in kilowatts. The dotted line represents the requirement used in the cost estimation.

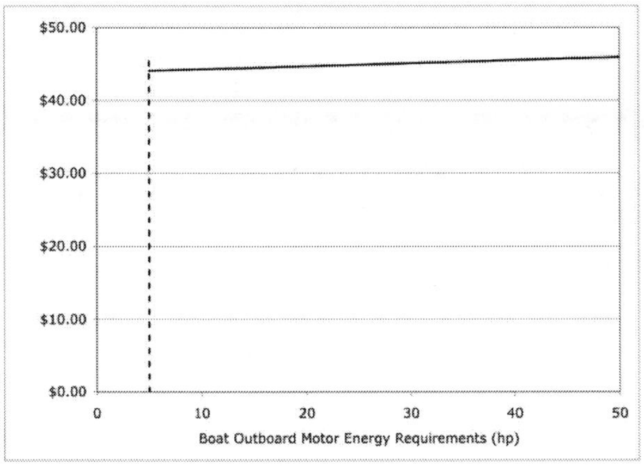

Figure 4b: Biomass price per dry ton as a function of transport boat energy requirement in horsepower. The dotted line represents the requirement used in the cost estimation.

One final parameter that can be improved is the cost of electricity. A standard value of $0.15 per kilowatt-hour was used in the estimation. Electricity could potentially be generated onsite through the burning of lignin or residual biomass, further reducing the price of the biomass, as shown in Figure 5. Additionally, electricity could be obtained for less depending on the location of the plant and proximity to a power source.

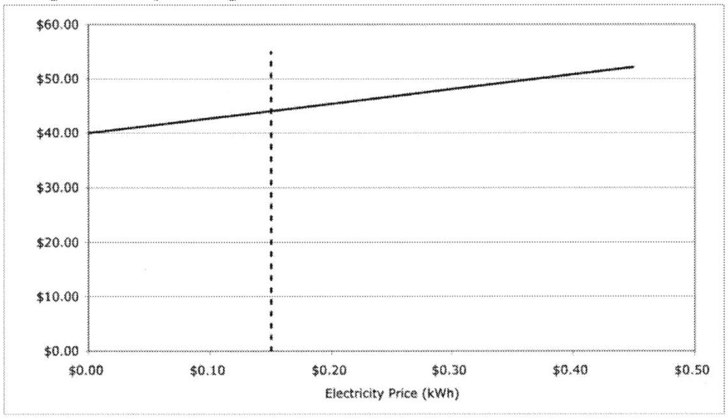

Figure 5: Biomass price per dry ton as a function of electricity kilowatt-hour price. The dotted line represents the cost used in the cost estimation.

Implementation of any one or a combination of these improvements could make water hyacinth a cost effective choice as a biomass feedstock.

Conclusions

While there are many biomass feedstocks available for sustainable fuel production, the most desirable feedstocks will possess certain traits. Plentiful availability, a low cost, frequent harvesting times, and a low lignin content are all advantageous traits which water hyacinth possesses. Additionally, the high water and low lignin contents of the plant make it particularly amenable to partial digestion, which can be used as an inexpensive pretreatment strategy. Our preliminary economic life cycle analysis has shown that the combined growth and digestion properties of *E. Crassipes* make it an economically competitive source of biomass, at approximately $44 per dry ton. A sensitivity analysis was also performed to identify parameters (such as cut width, speed, and power requirements) that could be perturbed to further decrease the cost of production. Improvements in these factors could potentially reduce the price of biomass down into the $30 per dry ton range. Water hyacinth cultivation and harvesting could also stimulate our nation's rural agricultural economy, while decreasing our dependence on foreign oil.

References

1. Solomon, B. D., Barnes, J. R. & Halvorsen, K. E. (2007) *Biomass and Bioenergy* **31**, 416-425.
2. (2006) *Breaking the Biological Barriers to Cellulosic Ethanol: A Joint Research Agenda DOE/SC-0095* .
3. Dias De Oliveira, M. E., Vaughan, B. E. & Rykiel, E. J. (2005) *BioScience* **55**, 593-602.
4. Chen, M., Xia, L. & Xue, P. (2007) *International Biodeterioration and Biodegradation* **59**, 85-89.
5. EERE *Biomass Feedstocks, U.S. Department of Energy, Energy Efficiency and Renewable Energy* .
6. Ulrich, D. & Vasudevan, T. .. (2004) in *Chemical Engineering Process Design and Economics: A Practical Guide*. (Process (Ulrich) Publishing, Durham, NH), pp. 409-435.
7. Petrell, R. J. A. B., Smerage, L. O. A. & H., G. (1991) *Journal Of Aquatic Plant Management* **29**, 45-50.
8. Penfound, W. T. A. E. & T., T. (1948) *Ecological Monographs* **18**, 447-472.
9. Marine, L. W. W. & Ltd., E. C. (2007) *Liverpool Water Witch Waterway Maintenance Solutions* .
10. (1997) *Strangers in paradise: Impact and Management of Nonindigenous Species in Florida*, eds. Simberloff, Schmitz, C. & Brown, C. (Island Press, Washington DC).
11. Turhollow (1994) *Biomass & Bioenergy* **6**, 229-241.

454d Supercritical CO2 Extraction Lipids And Phospholipids From Sewage Sludge

Kaiwen Liang, Hossein Toghiani, and Rafael Hernandez. Chemical Engineering, Mississippi State University, 330 Swalm Circle President Dr., Mississippi State University, MS 39762

Supercritical CO2 extraction was used to fractionate and enrich high value phospholipids from sewage sludge. Lecithin was selected as a model to evaluate supercritical CO2 extraction lipids and phospholipids from swage sludge. Neat supercritical CO2 was used to extract oil from soybean lecithin and then supercritical CO2 with 5 – 15 wt% methanol as a co-solvent was used to extract phospholipids from de-oil soybean lecithin were performed. Gas chromatography (GC) fitted with a flame ionization detector (FID) was used to quantitatively analysis the fatty acid composition of extract. Fourier transform infrared spectroscopy (FTIR)-attenuated total reflectance (ATR) and nuclear magnetic resonance (NMR) was used to qualitatively and semi-quantitatively analysis lipids and phospholipids in the extract. Neat CO2 was used to extract the oil from sewage sludge. The sludge was then extracted with supercritical CO2 over a range of temperatures, with 5 – 15 wt% methanol as a cosolvent. Phosphorus analysis was also performed to quantitatively analysis phosphorus content in the sample and extracts. The effect of extraction temperature, level of co-solvent employed and extraction pressure were examined.

454e Integrated Environment for Detailed Modelling and Evaluation of Gasification Processes for Energy Production

Mar Perez-Fortes, Sergio Ferrer-Nadal, Aaron D. Bojarski, Georgios Kopanos, Jose M Nougués, Enric Velo, and **Luis Puigjaner**. Chemical Engineering Department, Universitat Politècnica de Catalunya - ETSEIB, Diagonal, 647, Barcelona, E-08028, Spain

Gasification is becoming one of the best clean technologies for power production. The produced gas from gasification of coal and/or other solid wastes may be utilized in gas turbines of Integrated Gasification Combined Cycle (IGCC) systems and fuel cells. However, such systems are quite complex to model and usually lack of detailed integrated models that could improve their flexibility and performance. Within the different types of gasifiers contemplated in IGCC systems, entrained bed reactors offer an excellent gas-solid contact leading to very good heat transfer and highest gasification rates.

In this work, a model of an entrained bed gasifier has been developed for improved synthesis gas production. The gasification model considers five stages: pyrolysis, volatiles combustion, char combustion, gasification and a final gas equilibrium zone. Energy integration within the reactor section has been carried out by accounting all the energy consumed and generated by the process. Furthermore, this model is envisaged to evaluate energy losses as well as external heat supplies or refrigeration requirements in order to keep a steady reaction operation.

Subsequent raw gas cleaning operations have also been modelled. They comprise fly ashes, ammonia, sulphur, and metals removal. Other indispensable operations are also included to minimize the environmental effects of the effluents of the gasification plant: a waste water stripping unit and a modified Claus plant for sulphur recovery.

Aspen Hysys (AH) is chosen as the platform for the overall process simulation of the entrained bed gasifier because of its capability for developing custom unit operation models as extensions, which become part of the simulation as any other AH object. Available AH models have been used for conventional process operations, while custom models have been developed for more specific stages within the gasification process, as pyrolysis, char combustion and gasification reactions; also in gas purification units, as the COS hydrolysis reactor and the Claus plant.

Thorough simulations were necessary to help finding out feasible operating conditions of the process to achieve better process performance. The solution adopted is able to predict the composition of the produced gas under various operating conditions (temperature, fuel/carburant ratio, etc.) and from different fuels mixtures (coal, wastes mainly, and biomass in limited proportions).

ELCOGAS data has been used in the platform testing and validation studies. The proposed gasification plant model improves the understanding of the whole process, alternative process synthesis assessment and can be also used as an accurate predictive tool at the optimization stage towards an optimum design.

Acknowledgements:

Financial support received from the European Community projects (RFC-CR-04006), the Generalitat de Catalunya with the European Social Fund and the Ministerio de Educación y Ciencia is fully appreciated. ELCOGAS provision of data is also acknowledged with thanks.

Enhanced Production of Liquid Fuels by using a Syngas Chemical Looping Process

Luis G. Velazquez-Vargas, Fanxing Li, Deepak Sirdar, Ray Kim, and L.-S. Fan.
The Ohio State University
Department of Chemical and Biomolecular Engineering
140 W 19th Ave,
Columbus, OH 43210

1 Introduction

Due to the inflated petroleum and natural gas price, the synthesis of liquid fuels from coal via the Fischer-Tropsch (FT) synthesis has received large interest. In this process, liquid and gaseous hydrocarbons and oxygenates are produced by the catalytic hydrogenation of CO in a bubbling bed reactor. Typically, the H_2/CO molar ratio of the FT reactor feedstock is around 2, which is much higher compared to the ratio obtained from the syngas of modern gasifiers such as GE (~0.67) or Shell (~0.5) gasifiers. Currently, most of the hydrogen is produced by natural gas reforming. However, hydrogen production from coal is potentially a more economic alternative due to increasing natural gas prices. Current hydrogen production technologies use direct coal gasification to produce syngas, which is then "shifted" to produce more hydrogen. In the water-gas-shift operation, the CO contained in the syngas reacts catalytically with steam to form CO_2 and H_2. This reaction can be seen as

$$CO + H_2O \rightarrow CO_2 + H_2 \qquad (1)$$

Lower temperatures favor the thermodynamic equilibrium of the reaction, but also slow the rate of reaction. Thus, a catalyst is required to enhance kinetics. Usually the WGS operation is performed in two stages. First, the reaction is run at temperatures ranging from 350 °C to 450 °C using an iron based catalyst. This operation converts bulk quantities of CO. The second stage converts the remaining CO and is performed at lower temperatures which range from 200 °C to 300 °C (Kent 1974; Ruettinger, 2003). One drawback of the water gas shift reaction is that the H_2 gas contains large amounts of CO_2, which must be separated and removed if high purities of H_2 are required. There are various purification technologies, including metal membranes, polymer membranes, and absorption methods that can be used at large scales (Han, 2002). However, these methods are costly and reduce the overall energy efficiency of the system. The CO_2 produced is usually obtained at atmospheric pressures. The CO_2 has to be compressed for its transportation to a sequestration site. It is estimated that 75% of the total energy required for CO_2 management comes from separation, while the remaining is consumed by CO_2 compression and transportation (Mattisson, 2001). If a carbon dioxide policy is implemented, coal to hydrogen technologies should reduce the cost associated with CO_2 separation and transportation.

A novel patented process, the Syn Gas Redox (SGR) process, has been developed by OSU, and can convert coal derived syngas to H_2 at very high efficiency (Fan et al., 2007).

This study discusses the synergy between the SGR system and the FT process for the production of liquid fuels. The hydrogen stream from the SCL is used to increase the H_2/CO ratio from the gasifier outlet. The inherent CO_2 separation in the SGR system requires no additional CO_2 separation processes. Unconverted FT products are processed in the SGR system to produce H_2 improved the system selectivity towards a single product. As a result, improved system efficiency as well as selectivity towards liquid fuels can be achieved. Results show that the combination of a SGR looping with a FT process can enhance the production of liquid fuels such as diesel up to 30%.

2 Process description

A novel process has been developed to circumvent the shortcomings of the WGS reaction, namely the temperature swings and CO_2 separation. The SGR process produces H_2 from syngas at high temperatures with an integrated, high-pressure CO_2 production. Figure 1 shows a simplified schematic representation of the process. This process involves three operations: A) CO_2 production, B) H_2 production, and C) particle regeneration. In the CO_2 production operation, the reductive synthesis gas reduces an iron oxide particle to metallic iron.

$$3CO + Fe_2O_3 \rightarrow 2\,Fe + 3\,CO_2 \qquad (2)$$
$$3\,H_2 + Fe_2O_3 \rightarrow 2\,Fe + 3\,H_2O \qquad (3)$$

The reactions take place at high temperatures above 600°C. The reaction products are carbon dioxide and steam. Since the only pressure drop that occurs is a result of friction losses in the reactor, these losses are negligible compared to the total pressure of syngas, which may be as high as 30 atmospheres (~450 psi). The stream, after an isobaric water condensation, will produce a high-purity, high-pressure CO_2 gas stream ready for sequestration. The sequestrable carbon dioxide stream is expected to be at pipeline pressures. Both reactions 2 and 3 have equal number of gaseous species on either side of the reactions. Thus, the thermodynamic equilibrium is not affected by the pressure of the system. As an additional benefit, high pressures increase reaction rates.

In the hydrogen production operation, the reduced metal is put into contact with high-pressure steam. The metal will react with the steam to produce H_2.

$$3\,Fe + 4\,H_2O \rightarrow 4\,H_2 + Fe_3O_4 \qquad (4)$$

This well-known reaction generates hydrogen from hot iron and steam, and will generate a relatively pure H_2 stream that may require some post-polishing in order to achieve higher purities. However, the size of this purification equipment will be greatly reduced in comparison to current gas separation schemes due to the low contaminant levels of the hydrogen being produced. The iron, however, oxidizes to magnetite (Fe_3O_4). Hence, a minor regeneration step with oxygen may be required.

$$2\,Fe_3O_4 + \tfrac{1}{2}\,O_2 \rightarrow 3\,Fe_2O_3 \qquad (5)$$

This reaction regenerates the metal oxide for reuse in the CO_2 production operation and produces heat that generates steam to be used in the hydrogen production operation.

The process can be implemented at the production scale using two or three similar packed bed reactors: one is in CO_2 production mode; one is in H_2 production mode; and the last is in H_2 regeneration mode. Once breakthroughs are seen, the gases can be easily switched from one reactor to the other in order to provide continuous operation. The advantage of the process proposed here, compared with the traditional WGS reaction, is that separate high-pressure streams of CO_2 and H_2 can be produced at high temperatures.

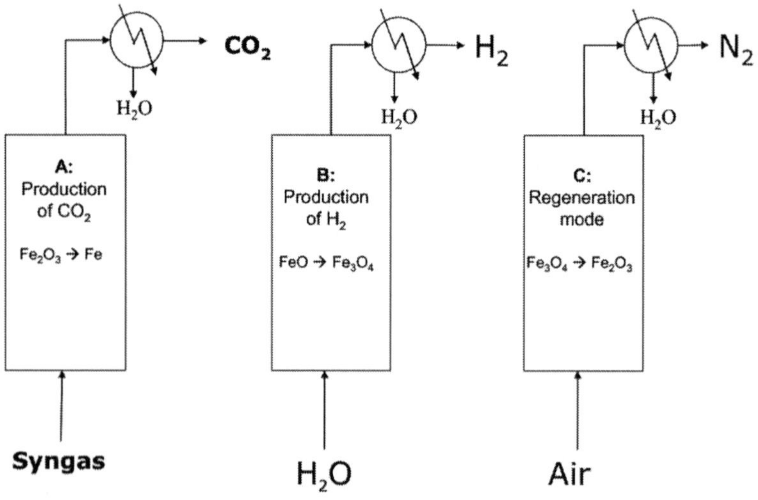

Figure 1. Schematic representation of the SGR operation process: A) CO_2 production and B) H_2 production.

3 Process Efficiency

A more comprehensive heat integration analysis was performed using ASPEN. Figure 2 illustrates the ASPEN analysis, with the following assumptions implicit in the calculations:

- Pittsburg #8 coal characteristics were used. A 1kg/hr coal basis was used.
- The reactor designs allow near-complete to complete conversions in Reactors A & B.
- Reactors A & B are adiabatic.
- Heat losses from exposed surfaces are not taken into account. This may correspond to approximately a 5% efficiency loss.
- All hot product streams are cooled from 900°C to 120°C to utilize their sensible heat to generate steam.
- CO_2 compression was not taken into account. This was done in order to compare this process's efficiency with the reported efficiencies of existing technologies which also do not take this energy into account. Based on our simulations, with CO_2 compression, the estimated drop in reported efficiency will be in the range of 7-12%

for all systems including the reported values for existing technologies.

- High pressure steam can be converted to electricity/work at 40% efficiency in a steam turbine.
- If the hydrogen requires compression from atmospheric pressure to 165 atm for storage, it is estimated that the drop in efficiency will be of the order of 5%. However, H_2 delivery pressure can be suitably increased by increasing the pressure of steam input to Reactor B. Thus, additional compression costs may be negligible or nonexistent depending on the final delivery option for hydrogen. For the calculations reported here, the energy requirement for hydrogen compression was assumed to be negligible.
- ASPEN component property models are representative of actual behavior.

The H_2 production efficiency is defined here as the Higher Heating Value (HHV) of the hydrogen produced divided by the HHV of the raw coal input. The sensible high grade heat from the various exhaust streams has been used to generate steam, as is the case with Reactor B. The excess steam produced is used in a steam turbine to run the air separation unit and the H_2 compressor. With these considerations, it is possible to make 0.148 kg H_2/ kg coal which corresponds to an energy conversion efficiency of 74%. Assuming heat losses and parasitic energy consumption to be of the order of 5%, the CLR efficiency is expected to be closer to 70%. This is substantially higher than the 64% hydrogen production efficiency of a coal gasification-WGS plant (Simbeck et al. 2002; Stiegel et al. 2006).

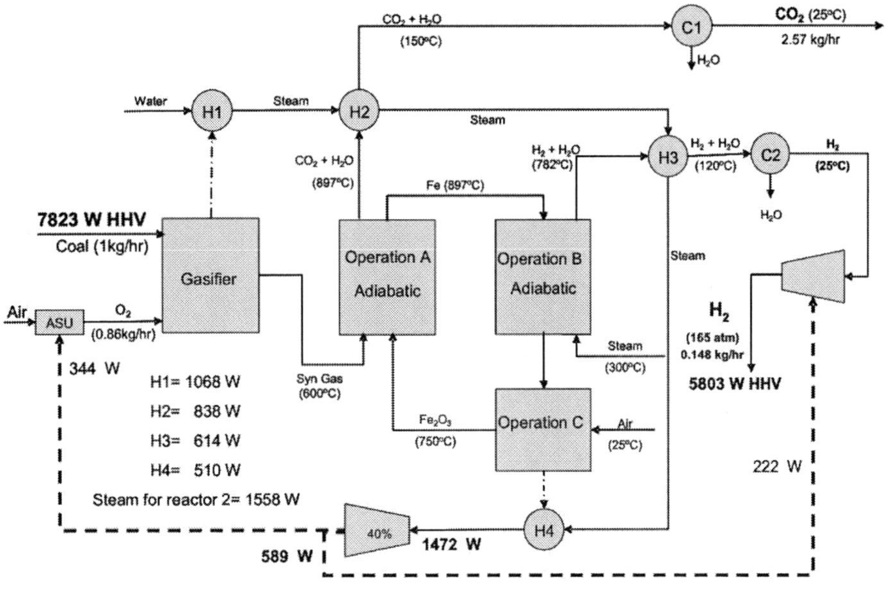

H_2 production efficiency = 74.2%

Figure 2 Heat integration and hydrogen production efficiency for the SGR process.

4. Integration of the SGR with a Fischer-Tropsch process

There are many different configurations to incorporate SGR process into the FT system. In one conservative configuration, the SGR process is used as a retrofit to the traditional coal-to-liquid plant. The role of SGR system is to compensate the hydrogen deficit in the untreated syngas. The SGR system, based on chemical looping concept, produces hydrogen by reduction and oxidation of iron oxide composite particles in a cyclic manner. In this process, gaseous fuel is introduced into a moving bed of iron oxide particles, where it is converted to carbon dioxide and water through the following reaction.

$$CxHyOz + (2x+y/2-z)MO \rightarrow (2x+y/2-z)Fe + xCO_2 + y/2H_2O \qquad (6)$$

Where MO refers to the different iron oxide phases. Reaction 6 reduces the iron oxide to metallic iron (M). The reduced iron particles are then introduced into a second moving bed reactor where they are reacted with steam to produce pure hydrogen and regenerate the iron oxide (reaction 7).

$$M + H_2O \rightarrow MO + H_2 \qquad (7)$$

In this process, the feedstock for SGR system is the gaseous products from the F-T reactor, and split stream of clean syngas. Figure 3 shows the flow diagram of the CTL-SCL process.

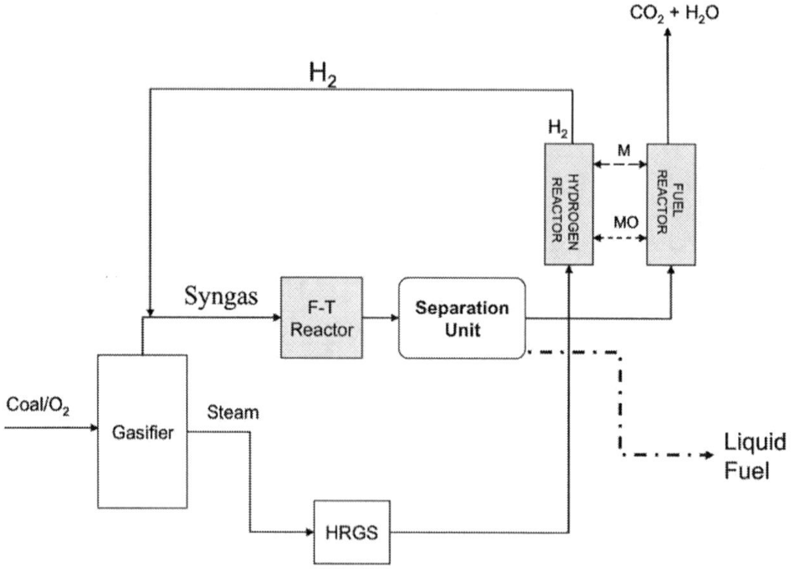

Figure 3. Block diagram of a Coal-To-Liquid plat integrated with a SGR system

According to mass and energy balances, this configuration can generate up to 30% more yield in liquid fuel production compared to traditional coal-to-liquid plants based on the water-gas-shift reaction. This increase in yield is due to byproducts of the FT reactor are converted in SGR system to produce H_2. Further, the steam needed for hydrogen production in SGR block is provided by gasifier and the F-T reactor.

5 Results and discussion

Fixed Bed reactor

A fixed bed reactor was constructed to demonstrate the hydrogen production from syngas using chemical looping. The overall diagram of the reactor is shown in Figure 4 The reactor is made using a quartz tube with a 16mm OD and 12mm ID. Stainless steel unions with Teflon fittings were used on both ends of the quartz tube to seal the reactor. A stainless steel capillary tube with a less than 0.5 mm inner diameter is used to deliver water from a Harvard PHD 2000 syringe pump. The capillary tube is inserted all the way to the heating zone of the reactor in order to serve as a steam nozzle. The small inner diameter of the capillary tube and the high temperature (~900 °C) at which the nozzle stays would provide stability in steam delivery and reduce pressure fluctuations due to the instability of the vaporization of superheated water. A gas inlet was installed with a diameter of 6 mm ID and joined to the tubing that holds the capillary tube with a tee. The gas would flow directly into the reactor outside of the capillary tube in any experiment, whether water is delivered to the system or not. The outlet of the reactor is connected to a Star 3600CX GC with a FPD detector dedicated for H_2S analysis. It is also connected to a PureAire STX-PA hydrogen analyzer and a Varian CP 4900 MicroGC for product gas analysis. A desiccant bed is installed between the reactor outlet and analysis equipment to remove the moisture from the gas stream. The heating zone is made up of a steel tube with an inner diameter of ~20 mm, which is only slightly larger than the outer diameter of the reactor tube. Heating elements and thermocouples were installed outside of the steel tube to heat the reactor and control the temperature between the inner wall of the tube and the surface of the quartz reactor. The overall length of the heating zone is about 600 mm and can be heated up to 1000 °C. The upper part of the heating zone (~100 mm) is always left empty to allow preheating of the steam and inlet gas. The reactor has a reasonably large inner diameter (12 mm) and effective length (500 mm). Therefore, it is capable of integral bed experimental studies, and the results from the fixed bed reactor would have direct implications towards the technology.

Table 1. Syngas composition for particle reduction in a packed bed (balance N_2)

Gas	H_2	CO	CO_2	H_2S
Concentration, %	23.99	44.75	5.32	10.37 (ppm)

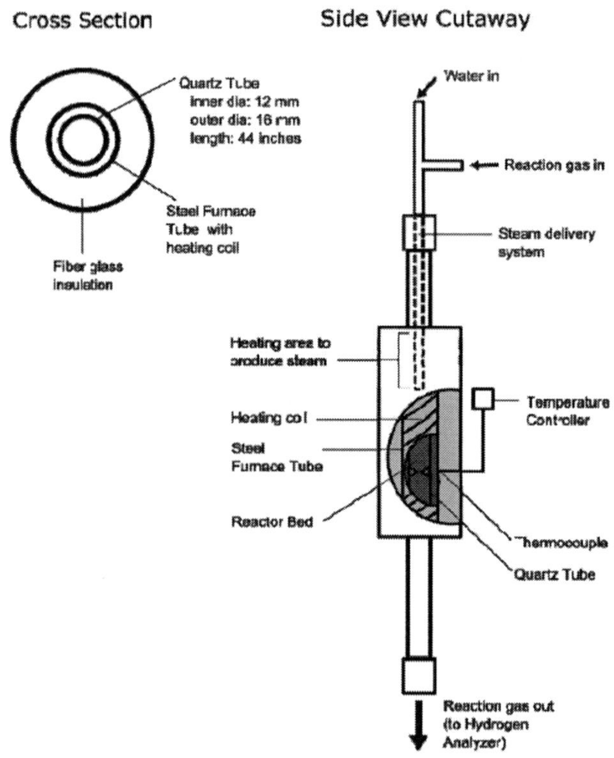

Figure 4. Schematic diagram of the fixed bed reactor

Reduction or iron oxide pellets with syngas

Experiments were carried out to study the reactions between the Fe-Ti-O particles and syngas. The gas composition used is shown in Table 1, which is similar to that of a Shell® type gasifier (Entrained Flow Dry Feed). Small amounts of H_2S were also sent into the system in order to simulate the H_2S concentration coming out of the hot gas cleanup unit. N_2 gas was used as a balance gas. About 28 g of pellets with a 5 mm diameter were loaded in a quartz tube, which corresponded to a bed height of approximately 19.5 cm.

The presence of CO_2 inhibits the reverse Boudaurd reaction and thus prevents carbon deposition. Thermodynamics predicts that a CO_2/CO ratio of ~1:20 or higher would prevent carbon formation under 900°C. This ratio should also be lower than 1:3 in order to fully reduce Fe_2O_3 to its metallic form. The CO_2/CO ratio in the syngas produced by Shell® gasifier is around 1:8.5, which falls squarely in this range. Therefore, no CO_2 is added in this experiment to prevent carbon deposition.

Figure 5 shows the CO, CO_2, and H_2 concentrations at the inlet (time < 0 min) and at the outlet (time > 0 min) of the reactor as a function of time. The pellets were considered to be fully reduced when the gas concentrations at the outlet were similar to those of the inlet. As can be seen from Figure 5, the breakthrough occurred within a very short period of time (less than 5 min). During this period, both CO and H_2 were fully converted, and the only gas detected from the outlet was CO_2. This validates the proposition that a high purity CO_2 stream is expected from the reactor. The short breakthrough time is due to the high syngas concentration and flow rate used in the experiment, which leads to a very fast conversion of the Fe-Ti-O particles. After the breakthrough, the gas concentration changes rapidly to that of the inlet concentration. The relatively large period of time (~15 min) required for the gas concentration to stabilize may be due to the large volume of the desiccant bed that is installed upstream of the MicroGC. This desiccant bed may absorb part of the gas.

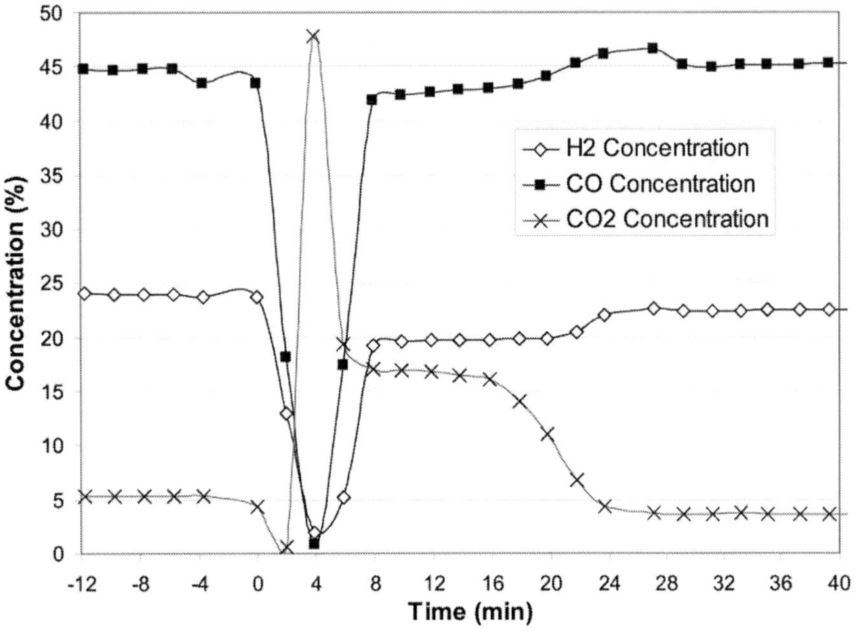

Figure 5 Breakthrough curve for the reduction of iron oxide pellets with syngas.

Oxidation with steam

Following the reduction of Fe-Ti-O particles using syngas, steam was introduced into the reactor system to produce H_2. This served two purposes: to assess the purity of the hydrogen gas and to corroborate minimal carbon deposition on the particles during reduction.

The value in assessing hydrogen purity is straightforward by the values given by the GC analysis. The second point concerning carbon deposition is of interest due to the possibility of the carbon reacting with steam to form CO through the carbon-steam reaction:

$$C + H_2O \rightarrow CO + H_2 \quad (8)$$

The packed bed experiment was performed by introducing a mixture of about 20% steam balanced with N_2 into the fixed bed reactor at 900 °C. The reactor was filled with reduced Fe-Ti-O particles (previously reduced with syngas). Figure 6 shows the concentrations of H_2 and CO at the outlet of the reactor as a function of time. The CO_2 concentration was found to be negligible. As can be seen, the steam was steadily converted into H_2. The fluctuation observed in the hydrogen concentrations was due to the unstable vaporization of water, which was injected using a syringe pump. An average of 16.5% H_2 concentration was obtained during the first 30 minutes of the experiment, which corresponds to about 82.5% steam-to-H_2 conversion. The breakthrough occurred after 30 minutes, and the hydrogen concentration steadily decreased for about 20-25 minutes. Figure 6 also shows very low CO concentration. The maximum CO concentration detected was about 0.05%, which suggests that little or no carbon deposition occurred during the reduction process. Based on the low concentration of CO observed, we can conclude that the presence of CO_2 during the particle reduction effectively inhibits the reverse Boudaurd experiment. This agrees with the previous findings in the TGA.

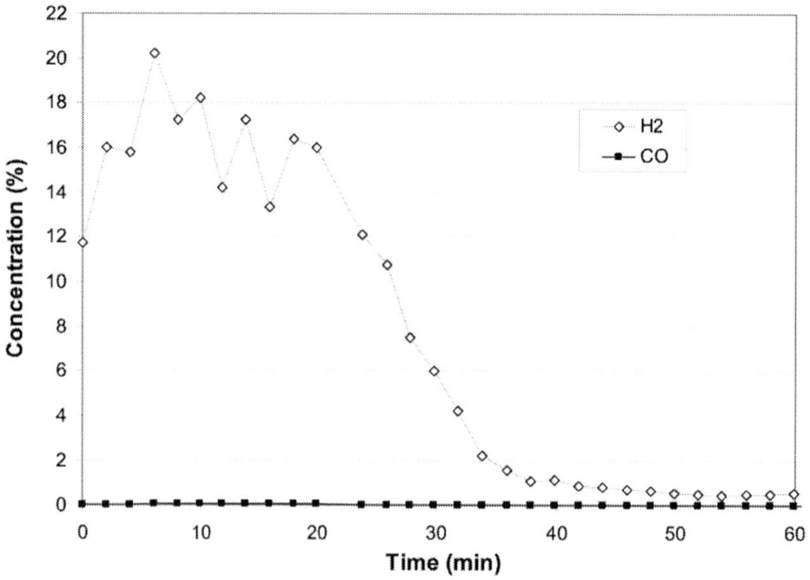

Figure 6 Breakthrough curve for the oxidation with 20 % steam of previously reduced pellets

6 Conclusions

There is substantial investment in researching coal gasification technologies. For example, new WGS reactors are being investigated in which H_2 can be separated *in situ* to drive the reaction further towards the products side. Research is also focused on the development of new catalysts which are less sulfur sensitive. However, post-gasification systems such as the SGR process may provide better hydrogen production efficiencies from syngas. The SGR process for hydrogen production replaces a conventional WGS reaction system by a chemical looping system. The process uses novel metal oxide composite particles that undergo reduction reaction with syngas to form reduced metal while yielding a high-pressure, sequestration-ready carbon dioxide stream. The reduced metallic form is subsequently oxidized with steam, forming hydrogen. The process has several advantages over a conventional WGS reaction system. It eliminates the costly and energy intensive carbon dioxide separation units. It also eliminates the need to perform a WGS shift reaction and a combustion of purge gases in the turbine/generator.

As a result, the hydrogen can be produced at efficiencies as high as 74%. This novel technology is readily adaptable for coal gasification units.

Experiments in packed bed columns demonstrate that this technology can deliver a pure hydrogen stream separated from carbon dioxide, thus circumventing the need for carbon dioxide separation methods. Further, this technology has the potential to reduce current syngas cleaning operations in gasification plants by avoiding the reaction of H_2S with iron oxide. The sulfur in the SGR process may be released along with the carbon dioxide gas, which can then be sequestrated. Moreover, When the SGR process is implemented in coal-to-liquid fuel plants, converts FT gaseous products such as methane or light hydrocarbon gases to Hydrogen, increasing the overall liquid fuel production.

7 Acknowledgements

Authors would like to acknowledge the Ohio Coal Development Office of the Ohio Air Quality Development Authority and United States Air Force for financial support for this project.

8 References

Fan, L.S., Gupta, P., Velazquez Vargas, L. G., Li, F. Systems and methods of converting fuel. PCT Int. Appl. 2007:788506(2007)

Han, J.; Kim, I.-S.; Choi, K.-S., “High purity hydrogen generator for on-site hydrogen production”. International Journal of Hydrogen Energy. 27, 10, 1043-1047, 2002.

Kent J. A. (Editor). “Riegel's handbook of industrial chemistry”, New York; Van Nostrand Reinhold,1974.

Mattisson. T. Lyngfelt, A. Cho, P., “The use of iron oxide as an oxygen carrier in chemical-looping combustion of methane with inherent separation of CO2”. Fuel. 80, 1953-1962, 2001.

National Academy of Engineering, “The Hydrogen Economy: Opportunities, Costs, Barriers, and R&D Needs”, Washington D.C.: The National Academies Press, (2004)

NETL report PED-IGCC-98-001, “Texaco Gasifier IGCC Base Cases”, June 2000.

Ruettinger, W.; Ilinich, O.; Farrauto, R. J. “A new generation of water gas shift catalyst for fuel cell applications” Journal of Power Sources. 118, 61-65, 2003.

Simbeck, D., Chang, E. Hydrogen Supply: Cost Estimate for Hydrogen Pathways – Scoping Analysis, USDOE NREL/SR-540-32525. July, 2002.

Stiegel G.J., Ramezan M. Hydrogen from Coal Gasification: An Economical Pathway to a Sustainable Energy Future. International Journal of Coal Geology 2006; 65: 173-190.

454g Integrated One Stage "Calcium Looping Process" For Enhanced Hydrogen Production And Liquid Fuel Synthesis

Shwetha Ramkumar, Mahesh V. Iyer, FuChen Yu, and L. S. Fan. Chemical and Biomolecular Engineering, The Ohio State University, 140 W. 19th Ave, 125A Koffolt Labs, Columbus, OH 43210

The total world energy demands for transportation, industry, electricity, etc are projected to increase from 420 quadrillion BTU in 2003 to well over 720 quadrillion BTU by the year 2030. Fossil fuels which include petroleum, natural gas, and coal continue to meet these demands. However, with the escalating crude oil and natural gas costs, the focus on the development of alternative technologies for the production of liquid fuels, hydrogen and electricity with high process efficiencies is gaining significance.

Clean coal technologies that include CO2 capture during the production of electric power, liquid fuels and hydrogen is one major thrust area. Coal gasification provides options for the conversion of syngas to liquid fuels, hydrogen or efficient power via the IGCC routes. Enhancement in the production of high purity hydrogen is limited by the thermodynamics of the water-gas shift reaction (WGSR). However, this constraint can be overcome by concurrent WGSR and carbonation (of calcium oxide) reaction to enhance H2 production. The carbonation of calcium oxide forming calcium carbonate incessantly drives the equilibrium-limited WGSR forward by removing the carbon dioxide (CO2) product from the reaction mixture. Calcium carbonate can be separately calcined to yield a pure CO2 stream for its subsequent sequestration and the calcium oxide recycled back. This calcium looping scheme not only improves the hydrogen yield and purity but also integrates a gas cleanup and CO2 management scheme in the hydrogen production process.

The proposed calcium looping scheme simplifies high purity H2 production by integrating the water gas shift reaction (WGSR) with in-situ CO2 and sulfur removal from the synthesis gas at high temperatures in a single stage reactor while eliminating the need for WGS catalyst. We have identified a high reactivity OSU patented, mesoporous calcium oxide sorbent for the in-situ capture of CO2 and H2S. The morphological properties of our high reactivity calcium sorbent (PCC) can be tailored using surface modifiers to demonstrate a high CO2 capture capacity of about 70% by weight (~700g of CO2/kg-sorbent) and a high reactivity towards H2S. Life cycle testing of the sorbent over multiple cycles of carbonation-calcination reactions shows that PCC sorbent attains a capture capacity of 40-36 wt% over 50-100 cycles, which is significantly higher than most of the high temperature sorbents reported in literature. Experiments reveal the superior performance of the PCC sorbent over naturally occurring limestone for H2 production. The integrated reaction schemes were investigated over varying pressures (1-20 bar), temperatures (550-700C), steam/carbon ratios and sulfur compositions in the feed stream for catalytic and non-catalytic operations.

The calcium looping process highlights include (1) production of a 99+% high purity H2 stream at high temperatures and pressures (2) simultaneous removal of CO2 as well as sulfur impurities (3) production of a sequestrable CO_2 stream by sorbent regeneration (4) reduction in excess steam requirement (5) flexibility in carbon monoxide conversion to produce H_2:CO ratios of varying compositions while reducing sulfur to minimal levels in the product gas stream (ppm to ppb levels), suitable for fuels/ chemical synthesis through the Fischer Tropsch reaction.

AUTHOR INDEX

AUTHOR INDEX

American Institute of Chemical Engineers
3 Park Avenue
New York, New York 10016-5991

ISBN: 978-1-60560-012-3